7-17-72

EXPERIMENTAL ASTRONOMY

ASTROPHYSICS AND
SPACE SCIENCE LIBRARY

A SERIES OF BOOKS ON THE RECENT DEVELOPMENTS

OF SPACE SCIENCE AND OF GENERAL GEOPHYSICS AND ASTROPHYSICS

PUBLISHED IN CONNECTION WITH THE JOURNAL

SPACE SCIENCE REVIEWS

VOLUME 18

JEAN-CLAUDE PECKER

Professor at the Collège de France

EXPERIMENTAL ASTRONOMY

SPRINGER-VERLAG NEW YORK INC. / NEW YORK

D. REIDEL PUBLISHING COMPANY / DORDRECHT-HOLLAND

L'ASTRONOMIE EXPÉRIMENTALE

First published by Presses Universitaires de France, Paris

Translated from the French by Robert S. Kandel

SOLE-DISTRIBUTOR FOR NORTH AND SOUTH AMERICA

SPRINGER-VERLAG NEW YORK INC. / NEW YORK

Title Number 8–82008

Library of Congress Catalog Card Number 77-118378

Printed in The Netherlands by D. Reidel, Dordrecht

FOREWORD

Socrates knew all that was known by his contemporaries. But already in the Middle Ages it was becoming difficult for a single man to have a truly encyclopedic view of all human knowledge. It is true that Pico della Mirandola, Pius II, Leonardo da Vinci, and several other great minds were thoroughly in possession of considerable knowledge, and knew all that one could know, except no doubt for some techniques. The encyclopedists of the 18th century had to be content with an admirable survey: they could not go into details, and their work is a collective one, the specialized science of each collaborator compensating for the insufficiencies of the others. We know very well that our science of today is a science of specialists. Not only is it impossible for any one person to assimilate the totality of human knowledge, it is impossible even to know ones own discipline perfectly thoroughly. Each year the presses of science produce a frightening quantity of printed paper. Even in very limited fields, new journals are created every day, devoted to extremely specialized, often very narrowly defined subjects.

It is indeed evident that in a field whose scope extends well beyond astronomical or astrophysical research, it is materially impossible to be informed of everything, even with the richest of libraries at hand. It is also clear that progress in the field is rapid; as the pace of progress quickens, the perfectly human time factor involved in the author's documentation will become more and more serious for the reader. You can see what I am coming to: this book does not pretend to be encyclopedic – that would be impossible; nor can it be said to be totally up-to-date. When it finally appears in print, it will be even less so than today. Furthermore, except for certain points, one can hardly innovate: in such a book proofs and examples cannot be very original; as for the other authors whose works have been – quite deliberately – plundered, the author can only ask their pardon. In preparing some of the chapters, no attempt has been made to give a complete bibliography: only a thread, like Ariane's, through the crowded bookshelves of the labyrinths of our libraries. And so, perhaps a bit egotistically – since he is now out of the labyrinth, and back on solid groud – the author hopes that the young scientists who read this book will find a way to defend themselves against the minotaurs (and they are legion) that they will enconter there.

And – alas! – they will have to defend themselves against many other monsters too!

Serene and fertile reflection – what is left of it in overcrowded laboratories, with the racket of computers which are less our slaves than we are theirs, when we are faced with floods of data to reduce and requirements of ever higher accuracy? What is left of it in the face of the H-bomb, the flight to the Moon, famine in India, and war in

Viet-Nam? What is left of it in a world where the irreversible abolition of distances and barriers has made even more cruelly acute the solidarity of all man, no longer to be pushed out of our minds? Young scientists cannot forget that their science can be used for good, but also for ill. And those who enter space research should be even more aware of this than most.

Even so, it seemed to the author that in this little book* he could give the purely scientific reasons for space research – quite apart from military or political imperatives. Although this can only be an introduction, young research astronomers should be able to use it to justify their efforts in this discipline of space research – so powerfully attractive, for other reasons. It is true that, as far as astronomers are concerned, one cannot yet say that astronomy has learned very much from the utilization of rockets, satellites, and extra-terrestrial probes. However, the conquest of space will open enormous possibilities. A crumb of the funds devoted to it (for other reasons), if properly used, will be enough to multiply astronomical knowledge enormously. Could we hope that one day, scientific reasons alone will motivate astronautics and justify the necessary investments?

For the above reasons, and in order to stay strictly within the limits defined by the title of this book, I have deliberately omitted the technical side. There is nothing on propulsion, very little navigation, and just enough geophysics for an understanding of the problems of the astronomers. Above all I have tried to show what space technology, by its very existence, can give to astronomers, and also what the requirements of the astronomers might be. Obviously these possibilities and these requirements are closely linked to the progress of the technology: but I have not felt the need to mention this more than incidentally: the necessary references can be found in abundance in the bibliographical appendix.

You might then ask – 'Why write such a book?'

Such a clear question deserves an answer. Of course, I won't say that this book translates an internal need (I prefer calm reflection and slow research) nor that it satisfies an external need (naturally, it costs money to build a summer house, but must we say so?). I will say that I have tried to introduce a personal point of view, if only in the organization of the book; I have tried to explain (without begging the question, as far as possible); I have tried to be provocative and so to force a certain amount of reflection on young scientist; I have also tried to underline those aspects of space research which I consider important (I may be wrong, but not very much, I think), and will not go out of fashion too quickly.

I have tried neither to teach, nor to substitute for a complete course; I have tried only to interest and to stimulate. I am quite aware that I have succeeded only partly in this: should the dissatisfied reader, his curiosity heightened by my contribution, go on

* And in another volume by the same author, *Les observatoires spatiaux* (Collection 'La Science Vivante'), P.U.F., Paris, 1969.

to read other works on the same subject in order to satisfy that curiosity, then at least I will be consoled for the defects of this book.*

Acknowledgments

The author would like to thank all those who helped him in the preparation of this monograph and in furnishing photographs, and especially Mr. R. Futaully, who prepared both Figure 1 and Tables I and II, and Miss G. Drouin, who edited the manuscript. Dr. F. Barlier was kind enough to read and comment on the manuscript before its completion.

I am deeply indebted to my translator, Professor R. S. Kandel, who has not only made a fine translation but also improved the original on several points, corrected a few mistakes, and made it when possible up-to-date.

---

* This work deals with two types of problems: extra-terrestrial devices, considered as astronomical objects (Chapters I, II, III, IV), and the direct exploration of the astronomical universe (Chapters V and VI); through space research, these change the nature of astronomical research completely. In a second work by the same author, *Les Observatoires spatiaux,* we shall deal with a third problem, namely the extension of classical astronomical techniques by space technology. These two works complement one another quite naturally.

TABLE OF CONTENTS

CHAPTER I

ASTRONOMY – AN EXPERIMENTAL SCIENCE?

1. Auguste Comte and Astronomy

From Aristarchus and Hipparchus, to Tycho Brahe and Galileo, Kepler and Newton ... To the philosophers of the 19th century, in their concern with the way (or ways) in which the human mind functioned, astronomy, in its double role as science of the natural universe and paradise of the abstract, appeared to be a privileged domain. Auguste Comte wrote:

Inasmuch as astronomical phenomena are the most general, the simplest and most abstract of all, natural philosophy must obviously commence by their study, since the laws governing them influence the laws governing all other phenomena, while remaining on the contrary essentially independent of the latter.

Further on he added:

It is indispensable to separate clearly celestial physics and terrestrial physics, and to proceed to the study of the latter only after that of the former, from which it derives.

In truth this was already quite questionable at the time of Auguste Comte; even so, only recently, someone could write (no doubt rather by some ironic nostalgy than by objective conviction, but nonetheless not without serious reasons) that astrophysics ...

might even include all of physics, inasmuch as the laboratory of astrophysics, the most complete and vast of any, includes and extends beyond terrestrial laboratories.

In any case, a second affirmation of Auguste Comte, symmetric to the first, was just as questionable. Astronomy was not only to be the ideal example of natural science, it was also the ideal field of application of mathematics, the abstract science *par excellence*:

There is perhaps not a single analytical procedure, not a single geometric or mechanical theorem, which is not today applied in astronomical research. This is why the science of celestial bodies should be placed at the very top of the hierarchy of our knowledge of nature.

This science also merits the first rank by the fundamental nature of the laws it studies. I have always considered it truly a stroke of philosophical genius on the part of Newton, to have given the title: *Philosophiae naturalis principia mathematica* (Mathematical principles of natural philosophy) to his admirable treatise on celestial mechanics. Thus he indicated, in the most energetic and concise way possible, that the general laws of celestial phenomena are the basis of the entire system of our real knowledge. Could our mind consider any terrestrial phenomenon in a truly scientific manner, without first considering the nature of this Earth in the universe of which we are a part, since its position and motion must necessarily exert a preponderant influence on all phenomena on it? What would become of our physical concepts, and thus our chemical and physiological concepts, without the fundamental notion of gravitation, which dominates them all?

And yet, while in this fine enthusiasm, Comte was unaware of (and in the excesses of his positivism, he preferred to remain so) the possible development of astronomy, and we know today to what extent this has been and will be tremendous! He even affirmed:

... This generality, this simplicity, and this abstract character result from the conditions in which we find ourselves placed with respect to the celestial bodies: all astronomical research must in the end be reducible to a visual observation. Thus, of all natural entities, the stars are those which we can study under the least variable conditions. We conceive of the possibility of determining their shapes; distance, size, and motions; on the other hand, there is no way by which we can ever hope to study their chemical composition, their mineralogical structure, nor – even less so – the nature of organized beings living on their surfaces. Our positive knowledge concerning celestial bodies is necessarily limited to their geometric and mechanical properties alone, and can in no way whatsoever comprehend the physical, chemical, physiological, and even social research which applies to those entities accessible to all our varied means of observation...

I hope that I may be excused for such massive quotations from Auguste Comte. But for a century – far too long! – his attitude defined what astronomy should have been in order to fit into the rigid framework of his system: and – alas – how many astronomers, especially in France, have continued the pursuit of an obsolete astronomy, under the more or less conscious impetus of the *a priori* notion that August Comte's argument was fully rational!

However, it does seem clear that, while it does not fill the place both too narrowly and too strictly established for it by Comte, astronomy, by its methods of acquiring information about the real universe, is indeed quite different from other sciences. It is not an *experimental* science, but rather a passive, *observational* science. Claude Bernard put it very well:

An astronomer first makes some observations, and then reasons about them in order to extract a set of ideas which he verifies by further observations made under appropriate conditions... Basically every science reasons in the same way and seeks the same goal. All of them seek a knowledge of the laws governing phenomena, so as to be able to predict, vary, or control these phenomena. Now while astronomy predicts celestial motions, and deduces from them a host of practical concepts, it cannot modify celestial phenomena by experimentation, in the way chemists and physicists can work in their science.

At the time of Claude Bernard, this also was (or seemed to be!) the case of biology. And so once again we have the lyric flights of Auguste Comte:

We might now ask why the science of astronomy seems more closely related to biology than to physics or chemistry. This is because, despite the indispensability of physics and chemistry, astronomy and biology, by their very nature, constitute the two principal branches of natural philosophy proper. These two great studies, the one complementary to the other, embrace in their rational harmony the general system of all our fundamental concepts. For the one, the Universe, for the other, Man: extremes, between which our real thoughts will always lie. First the Universe, then Man: in the purely theoretical order, such is the positive working of our intelligence; even though in the directly active order, it must necessarily be the inverse. For the laws of the universe dominate those of man, and are not modified by them. Between these two correlative poles of natural philosophy we find spontaneously inserted, on the one hand, the laws of physics as a sort of complement to astronomical laws, on the other hand, chemical laws, immediate preliminary of biological laws. Such is, from the most elevated philosophical point of view, the indissoluble rational spectrum of the various fundamental sciences.

The very fact that man could not possibly act on astronomical phenomena was what gave astronomy its greatness, in Comte's eyes:

This necessary preponderance of the science of astronomy in the first systematic propagation of the positivist approach is fully in agreement with the historical influence of this science, till now the principal motor of the great intellectual revolutions. Indeed, the basic feeling of the invariability of natural laws had to develop first with regard to the simplest and most general phenomena, whose superior regularity and grandeur display to us the only real order completely independent of any human modification.

Moreover Comte is certainly correct in affirming the historical influence of the development of knowledge of the heavens:

Even before it took on the slightest truly scientific character, this class of concepts was primordial in determining the decisive transition from fetishism to polytheism, which everywhere resulted from the cult of the stars. Afterwards, the first mathematical cosmology in the schools of Thales and of Pythagoras, constituted the principal psychological source of the decadence of polytheism and the rise of monotheism. Finally, the systematic growth of modern *positivism*, tending openly to a new philosophy, is essentially the result of the great renovation of astronomy initiated by Copernicus, Kepler, and Galileo.

Man unable to act on astronomical phenomena? What should we think of it? Biological experimentation became routine long ago. Well before Comte, how can the farmers of ancient times, or the Dutch growers of new tulips, be called passive observers? What about Leeuwenhoek's experiments on the viviparity of aphids? What about all surgeons, since Ambroise Paré? What about all the pharmacologists? And Indian priests of peyote and mescalin, experimenters in hallucinations? Lavoisier and respiration, Galvani and muscles? Claude Bernard himself! Without some experimentation, no medical progress would be possible.

Might it not be a bit the same for astronomy? Perhaps this strangely privileged position as the only non-experimental natural science does not belong to astronomy.

2. An Observer is also an Experimentalist

This question is legitimate. In any case no observation is truly passive. The data which after passing through various intermediaries may finally inform the observer about the observed, is considerable. The observer may select a part of the information – that which he needs to measure one quantity or another, whose value appears essential for him to know (right or wrong, and often only provisionally, in a given state of the science), either for its own sake or as a basis for further research.

Not only can the observer choose the observation, he can also adapt his instruments to the measurements he wants to make. He can repeat these measurements as many times as he judges necessary.

In fact, the great mass of possible data is (if one improves the instrumentation) an endless source of real experimentation. Let us take a simple and classical example, that of the determination of the temperature of the sun. With our astronomer, let us consider that such a measurement would have real 'value' as knowledge: this is the initial hypothesis.

The first operation for the astronomer is to read the existing literature on the subject, and to learn the techniques used to determine temperatures of various bodies in physics laboratories. Of course he will see immediately that there is no sort of thermometer he can use..., but he sees that the radiation emitted by heated bodies is related to their temperature by an experimental law (the black-body law) whose theoretical foundation is known since Planck. Thus the first idea of our observer: to measure the integrated intensity of the solar radiation (the 'solar constant'). The distance of the sun will yield the value of the 'effective temperature' of the star. This is the temperature of the black body emitting the same energy per square centimeter of surface as the sun itself.

We shall pass over the difficulties of measurement of the solar constant: we do indeed obtain a value for the effective temperature, $T_{eff} = 5807°$ (with accuracy $\pm 30°$).

However, the astronomer has good reasons for thinking that the sun does not have exactly the properties of a black body. And this is where experiment enters. If the sun were a black body, its 'color temperature' measured by comparing the intensity of its radiation at two wavelengths – e.g., blue and yellow – would be 5807°. Its 'brightness temperature' measured by determining accurately the absolute intensity of the radiation at a given wavelength, would also be equal to 5807°.

The 'experiment' to be carried out (and this is indeed experimentation) is easy to define: it is to construct a device which will enable us to determine brightness and color temperatures, choose wavelengths and make the measurements. And of course our astronomer finds that the temperature varies from one determination to the next. The results of the experiment are clear: the sun is not a black body, it is not infinitely opaque as is the black body of physics, nor is it isothermal: the variation of the temperatures found reflects both the variation of temperature with depth and that of opacity with wavelength.

Thus this first experiment leads to a more complicated interpretation: we must replace the notion of an opaque isothermal sun by that of an atmosphere; instead of a *single* 'temperature', we need a 'model'. But is the model which fits the measurements compatible with 'physics'? From one we are sent back to the other incessantly. The theory interprets the measurements, but physical reasoning shows that perhaps the theory will not be able to explain all of them: the measurements to be made must be judiciously chosen, better adapted instruments must be built: this is experimentation.

Contradictions appear. Two consequences: the physics of the solar atmosphere becomes clearer, the model becomes more complex. Today solar models include inhomogeneities, distributions of temperatures and pressures (at each depth), the description of velocity fields, magnetic fields... who knows?

3. Numerical Experimentation

In addition, there is another type of experimentation, very much used by astrophysicists. This is what might be called numerical experimentation. As the physical description of a celestial object (and let us take again the solar atmosphere as example) becomes more and more complex, very elaborate methods of analysis may become necessary. This is easy to understand. If a single temperature is sufficient to describe the atmosphere, a single measurement is needed; the relation between the measured quantity and the temperature sought is simple. If two temperatures suffice (that is if we can replace the model by a linear variation of the temperature as a function of some given parameter) the calculation of these temperatures from the measured flux is already difficult. Each measurement is the result of an integration involving the variations of the quantities sought as a function of the depth τ, the surface of exploration chosen on the solar disk σ, the wavelength λ, etc. Thus:

$$\text{The measured quantity} = \int_{\Delta\sigma} d\sigma \int_{\Delta\lambda} d\lambda \int_0^\infty \mathscr{F}(\sigma, \lambda, \tau)\, d\tau. \tag{1}$$

In principle each analysis is based on the inversion of such complicated integrals, and such inversions are always of limited accuracy and even to some extent *indeterminate*. Here $\mathscr{F}$ represents a function of *local* physical conditions, temperature, pressure, etc.

Faced with this inaccuracy and indeterminacy, astrophysicists have from the beginning been led to introduce theoretical 'models'. Stellar models, model atmospheres, sunspot models, models of cold planets, or of pulsating Cepheids... These are in fact artificial astronomical objects, which might exist in the sky, and which are capable of describing all the observed properties of the objects of which they are models. They appear as tables of numbers, abstraction of reality: a model is to the solar atmosphere a bit what the mathematical concept of a circle is to a 'circle' drawn on a sheet of paper. In order to construct a model, a complete coherent physical theory is needed, as well as a relatively small number of certain parameters – such as the *assumed* values of the affective temperature, gravity, or the chemical composition of the star or object to be described. Because of the indeterminacies mentioned above, the choice of these parameters is not unique. And so, with his computer, just like the chemist with his retort, the astronomer experiments. He chooses the stellar parameters; he computes a model, then the observable characteristics of the model and compares these to the measured ones. Sometimes by patient modelling, and subtle juggling, he is rapidly led to a satisfactory model. On the contrary, sometimes subtlety produces nothing; then he must use the brute-force method of constructing masses of models and choosing the best. Either way he will obtain a reasonable description of the star or object under study. Sometimes, however, he will not even succeed in that: we are far from the *indeterminacy* of Equation (1)! Instead the experiment shows that it is *impossible* to find any model that is good: the only recourse is to reject the theory – and to find a better one!

Furthermore, can physicists in fact experiment with the essential object of their study? Is it not, as with us, a question of choosing the techniques and the measurements to be made? One doesn't modify the laws of nature, to see what will happen! Does one act on the values of the fundamental constants of the universe? Does one modify the laws of attraction inside the atom? Just as astronomers, physicists experiment, choosing appropriately the measurements to be made, optimizing in a way the acquisition of information.

'Models' are not the tool of astronomers alone. There are models of atoms, models of molecules, models of plasmas... There are even models of a physical world in which the universal constants would have different values. Gamow carries this sort of experimentation to an extreme in the Mr. Tompkins series. Thus, in *Mr. Tompkins in Wonderland*, with a charming humor and for our enjoyment, he multiplies Planck's constant by... 10^{27}, thus reducing the speed of light to 15 km/hr. In this fantastic and charming experiment, billiards becomes a quantum sport, cyclists are flattened like pancakes, quantum gazelles diffuse through the bamboo, and the universe fits into a glass!

4. Conclusion: Space Astronomy Enters the Scene

Thus, it seems to me that from the philosophical point of view, there is no great difference between the methods of astronomy and those of other physical and natural sciences. Between the object of research and knowledge itself, scientific work deals with the necessary intermediaries. Experimentation is translation, for knowledge cannot be direct.

It is nonetheless true that the rockets developed in the Second World War opened a vast domain to astronomy, and that satellites and other extra-terrestrial devices allowed us to view astronomy under a truly new light, with the field of experimentation very much extended, and the possibility of eliminating some of the more frustrating intermediaries between the observer and the star under observation.

In this small book* we shall try to see in what way the development of space technology has raised hopes of broad development of astronomy, and to what extent these hopes have been satisfied. The 'intersection' of the two sets 'astronomy' and 'space research' is in fact still rather small in extent, but this extent increases regularly.

Quite naturally, we shall begin by speaking of the progress related to the fact that we now can experiment with artificial celestial bodies: satellites, meteors, comets – and so analyze the causes of their motions, and the structure of the upper atmosphere of the earth.

---

* And in the book *Les observatoires spatiaux* by the same author.

CHAPTER II

ARTIFICIAL SATELLITES AS CELESTIAL BODIES, OR THE INTRODUCTION OF 'EXPERIMENTAL CELESTIAL MECHANICS'

It would be vain to list all the extra-terrestrial or orbital devices launched since October 4, 1957, when the first Soviet Sputnik was launched. Tables I and II give the major launchings of astronomical interest since then, while Figure 1 shows the progress accomplished in terms of the number of shots and their efficiency. We shall limit the technical details to those given in these tables and this figure. For the following, let us suppose that the satellite has been launched; it has become an astronomical object. We can study it from terrestrial observatories, and radiotelescopes, theodolites, wide-angle cameras, etc., can follow its trajectory accurately. If the satellite emits no signals, or if it emits only a continuous signal at a well-defined frequency, it gives us no direct information on the medium in which it travels (geophysical medium: upper atmosphere; or astronomical medium: radiation field coming from all points of the universe, corpuscles of solar or galactic origin, interplanetary dust and meteoroids). In this case it is only through the study of its *movement* that we can extract information from this satellite about the various forces influencing

TABLE I[a]

Principal satellites of astronomical interest
(excluding space probes and sub-orbital shots)

Satellite	Launch date	Mission – Results
Sputnik 1	4 October 1957	First artificial satellite.
Sputnik 2	3 November 1957	Cosmic rays, UV and X-rays.
Explorer 1	1 February 1958	Cosmic rays and discovery of the first Van Allen belt.
Explorer 3	26 March 1958	Results complementing Explorer 1 and micrometeorite densities.
Sputnik 3	15 May 1958	Solar corpuscular radiation.
Explorer 7	13 October 1959	Solar and terrestrial IR, Ly-α and X-radiation; micrometeorite densities.
Transit 2A & NRL satellite	22 June 1960	Galactic noise measurements. Ly-α and X-rays.
Echo 1	12 August 1960	Mostly of geodesic and public interest.
Sputnik 5	19 August 1960	Cosmic rays and UV.
Discoverer 17	12 November 1960	Radiation in the internal Van Allen belt.
Explorer 11	27 April 1961	γ-ray astronomy.
Explorer 12	16 August 1961	Confirmation of the existence of plasma currents of solar origin.
Explorer 13	25 August 1961	Micrometeorites.
Transit 4B & Traac	15 November 1961	Determination of the figure of the earth. Internal Van Allen belt.

Table I (continued)

Satellite	Launch date	Mission – Results
OSO 1	7 March 1962	First 'orbiting solar observatory', solar (X, UV and γ) radiation, measurement of 'microflares' unobservable from the ground.
Cosmos 1	16 March 1962	Cosmic and solar radiation; micrometeorites.
Ariel 1	26 April 1962	Study of the ionosphere.
Explorer 16	16 December 1962	Micrometeorites: 15000 impacts in $7\frac{1}{2}$ months.
Hitch-hiker 1	27 June 1963	Van Allen belt measurements.
Vela 1 & 2 TRS 2	17 October 1963	Measurement of γ- and X-rays and of the neutrons in the solar and galactic radiation. Van Allen belts.
Explorer 18	27 November 1963	Measurement of the solar wind, discovery of a radiation zone beyond the Van Allen belts.
Echo 2	25 January 1964	Widely known passive satellite.
Electron 1 & 2	30 January 1964	Study of the Van Allen belts by two satellites.
Ariel 2	27 March 1964	Galactic noise, micrometeorites.
Electrons 3, 4	10 July 1964	Same as Electrons 1 & 2.
Explorer 23	6 November 1964	Micrometeorite detection.
OSO 2	3 February 1965	Solar studies.
Pegasus 1	16 February 1965	Meteoroid detection.
Greb 6	9 March 1965	Solar studies.
Pegasus 2	25 May 1965	Meteoroid detection.
Pegasus 3	30 July 1965	Meteoroid detection.
Molniya 1B	14 October 1965	First photographs of the earth taken from a distance of 36000 km.
GEOS-A	6 November 1965	Satellite constituting five geodesic experiments (including light emissions).
OAO 1	8 April 1966	First 'orbiting astronomical observatory', unable to carry out its mission because of a battery failure.
Pageos 1	23 June 1966	Geodesic satellite, similar in aspect to Echo, but much slower (angular motion: 5′ per second).
ATS 1	7 December 1966	Photographs of the earth from a distance of 36000 km.
OSO 3	8 March 1967	Same as OSO's 1 and 2.
OSO 4	18 October 1967	Fourth orbiting solar observatory.
ATS 3	5 November 1967	Color photographs of the earth from 36000 km distance.
GEOS-B	11 January 1968	Same mission as GEOS-A (whose optical signals ceased functioning on December 1, 1966).
Explorer 37	5 March 1968	Solar radiation studies.
Cosmos 215	18 April 1968	Astronomical satellite equipped with 8 telescopes (destroyed June 30, 1968).
ESRO 2B	17 May 1968	Study of solar and cosmic radiation.
Explorer 38	4 July 1968	Satellite for radioastronomical exploration.
Heos 1	5 December 1968	Study of solar and cosmic radiation outside the magnetosphere.
OAO 2	7 December 1968	As of January 6, 1969, the 11 telescopes of this second 'orbiting astronomical observatory' had accumulated 65 hours of observations (photographs of several hundred objects, including the planet Saturn).
OSO 5	22 January 1969	Same mission as OSO 4.
OSO 6	9 August 1969	Sixth orbiting solar observatory.

[a] This table, together with Table II and Figure 1, was prepared by Mr. R. Futaully.

Note concerning Tables I and II and Figure 1

Sources:

Satellites Situation Reports (Goddard Space Flight Center – NASA).
(Table of Artificial Earth Satellites (1957–1968), Table of Space Vehicles launched in 1957–1967 (Royal Aircraft Establishment).
TRW Space Log (TRW Systems).
Satellites et fusées porteuses lancés depuis Spoutnik 1 jusqu'au 20 juin 1965, by G. Nachszunow (8, allée du Docteur-Dupeyroux, 94-Créteil, France).

Remarks:

(1) The totals for 1966 do not take account of the three *objects discovered in orbit,* which, according to the Goddard Space Flight Center, "do not correspond to any known launching". The orbital elements are the following:

Designation	Period	Inclination	Apogee/Perigee
1966-00A	164.9 min	35.20°	6939/215 km
1966-00B	162.7 min	84.97°	6290/693 km
1966-00C	163.1 min	85.27°	6278/740 km

(2) It should also be noted that the number of launches is lower than the number of devices placed in orbit, since several satellites are sometimes placed in orbit by a single launch vehicle.

(3) As of 31 January 1969, only 950 of the 3600 objects placed in orbit since October 1957 were still turning around either the earth or the moon.

TABLE II[a]

Space probes and lunar or planetary shots

Name	Launch date	Mission – Results
Pioneer 1	11 October 1958	Reached the distance of 114000 km; measurements of the internal Van Allen belt, of micrometeorite densities in space and of the interplanetary magnetic field.
Pioneer 3	6 December 1958	Discovery of the second Van Allen belt.
Luna 1	2 January 1959	Passed within 6000 km of the moon, on January 4.
Pioneer 4	3 March 1959	Study of cosmic radiation around the moon (passed within 60000 km of the moon).
Luna 2	12 September 1959	First impact on the moon, September 13, 1959, at 21h 02m 23s UT.
Luna 3	4 October 1959	Passed within 6200 km of the moon on October 10, 1959. First photographs of the hidden face of the moon.
Pioneer 5	11 March 1960	Exploration of space between the earth and Venus.
Venus 1	12 February 1961	Measurement of cosmic radiation and magnetic fields around Venus.
Ranger 3	26 January 1962	Passed within 37000 km of the moon on January 28.
Ranger 4	23 April 1962	Impact on the moon (hidden face) on April 26, 1962, at 12h 40m UT.
Mariner 2	27 August 1962	Passed within 41000 km of Venus on December 14, 1962.
Ranger 5	18 October 1962	Passed behind the moon at 735 km distance.
Mars 1	1 November 1962	Passed about 200000 km from the planet Mars.
Luna 4	2 April 1963	Passed within 8500 km of the moon on April 6, 1962, at about 01h 26m.
Ranger 6	30 January 1964	Impact on the moon (Sea of Tranquillity), on February 2, 1964, at 09h 24m 33s.

Table II (continued)

Name	Launch date	Mission – Results
Zond 1	2 April 1964	Passed Venus at 100000 km in July 1964.
Ranger 7	28 July 1964	Impact on the moon (Sea of Clouds), on July 31, 1964, at 13h 25m 49s, after having transmitted 4316 photographs to the earth.
Mariner 3	5 November 1964	Study of the planet Mars.
Mariner 4	28 November 1964	Passed behind the planet Mars within 10000 km on July 15, 1965, and transmitted 21 photographs.
Zond 2	30 November 1964	Approached the planet Mars in August 1965.
Ranger 8	17 February 1965	Impact on the moon (Sea of Tranquillity) February 20, 1965, at 09h 57m, after transmitting 7137 photographs.
Ranger 9	21 March 1965	Lunar impact (Alphonsus cirque) on March 24, 1965, at 14h 08m 20s, after transmitting 6150 photographs.
Luna 5	9 May 1965	Lunar impact (Sea of Clouds) on May 12, 1965, at 19h 10m.
Luna 6	8 June 1965	Passed the moon at 160000 km.
Zond 3	18 July 1965	Approached and photographed the lunar surface at 9220 km distance.
Luna 7	4 October 1965	Lunar impact (Ocean of Storms), October 7, 1965, at 22h 08m 24s.
Venus 2	12 November 1965	Approached the planet Venus (24000 km) on February 27, 1966, at 02h 51m.
Venus 3	16 November 1965	First to reach the surface of Venus on March 1, 1966, at 06h 56m.
Luna 8	3 December 1965	Lunar impact (Ocean of Storms) on December 6, 1965, at 21h 51m 30s.
Pioneer 6	16 December 1965	Solar studies.
Luna 9	31 January 1966	Made a soft landing on the lunar surface and sent back numerous photographs.
Luna 10	31 March 1966	First artificial satellite of the moon.
Surveyor 1	30 May 1966	Made a soft landing on the moon (in the Flamsteed cirque), on June 2, 1966; transmitted more than 11000 photographs.
Lunar Orbiter 1	10 August 1966	Artificial satellite of the moon (took many photographs).
Pioneer 7	17 August 1966	Solar studies (same as Pioneer 6).
Luna 11	24 August 1966	Artificial satellite of the moon.
Surveyor 2	20 September 1966	Lunar impact September 23.
Luna 12	22 October 1966	Lunar satellite (photographs).
Lunar Orbiter 2	6 November 1966	Lunar satellite (photographs).
Luna 13	21 December 1966	Made a soft landing on the moon (Ocean of Storms) on December 24, 1966, and transmitted photos.
Lunar Orbiter 3	5 February 1967	Lunar satellite (took many photos).
Surveyor 3	17 April 1967	Soft landing on the moon (Ocean of Storms), April 20, 1967; had transmitted 6315 photographs by May 3, 1967.
Lunar Orbiter 4	4 May 1967	Lunar satellite; took 163 photos of the 212 planned for.
Venus 4	12 June 1967	Made a soft landing on the planet Venus on October 18, 1967, after studying the atmosphere.
Mariner 5	14 June 1967	Approached Venus at 3980 km on October 19, 1967.
Surveyor 4	14 July 1967	Crashed on the lunar surface July 17.

Table II (continued)

Name	Launch date	Mission – Results
Explorer 35	19 July 1967	Lunar satellite designed to study space between earth and the moon.
Lunar Orbiter 5	1 August 1967	Lunar satellite (photographs).
Surveyor 5	8 September 1967	Soft landing on the moon September 11, 1967; transmitted 18006 photographs by September 24, 1967; carried out the first chemical analysis of the lunar soil.
Surveyor 6	7 November 1967	Same mission as Surveyor 5 (moved 2 m as a result of a telecommanded jump).
Pioneer 8	13 December 1967	Same mission as Pioneers 6 and 7.
Surveyor 7	7 January 1968	Made a soft landing on the moon (20 km to the north of the crater Tycho) on January 10, 1968; transmits photographs and analyzes the lunar surface.
Zond 4	2 March 1968	Sent into solar orbit (parameters unknown).
Luna 14	7 April 1968	Satellite of the moon; study of the lunar gravitational field.
Zond 5	14 September 1968	Contained many living organisms (including several turtles); after having circumnavigated the moon, Zond 5 was recuperated in the Indian Ocean on September 21, 1968.
Pioneer 9	8 November 1968	For the prediction of solar flares (same mission as for Pioneers 6, 7 and 8).
Zond 6	10 November 1968	Flew around the moon on November 14, at a distance of 2420 km, and landed in the U.S.S.R. on November 17, 1968.
Apollo 8	21 December 1968	Three men take still and motion pictures of the moon from a circular lunar orbit at 112 km, on December 24, 1968.
Venus 5	5 January 1969	
Venus 6	10 January 1969	
Mariner 6	25 February 1969	Flew over the planet Mars on July 30, 1969, from west to east, transmitting 77 pictures of excellent quality.
Mariner 7	27 March 1969	Flew past the planet Mars on August 3, 1969, along a polar trajectory, transmitting 124 pictures.
Apollo 10	18 May 1969	Second flight around the moon by three men. The LEM makes two revolutions having a pericynthion of 15 km.
Apollo 11	16 July 1969	Third flight around the moon by three men. The LEM with two men on board lands in the Sea of Tranquillity on July 20, 1969.
Luna 15	13 July 1969	Crashed in the Sea of Crises on July 21, 1969, at 15h 50m 40s UT, after having gone through 52 revolutions about the moon.
Zond 7	7 August 1969	Same mission as Zond 6: flew around the moon on August 11, and was recuperated in the U.S.S.R. on August 14, 1969. Excellent photographs of the earth and the moon taken on August 8 and 11.

[a] See note to Table I, page 8.

this movement; we are dealing with celestial mechanics applied to objects launched by man in a deliberate way; we may, if we like, call this experimental celestial mechanics.

The forces acting on the motion of the satellite are evidently first of all *gravitational* forces; of all the gravitational attractions to which the artificial satellite is subjected, it is certainly, and by far, that of the earth which dominates. Other attractions, as well as non-gravitational forces, have only a perturbing effect on the motion of the satellite in the terrestrial field.

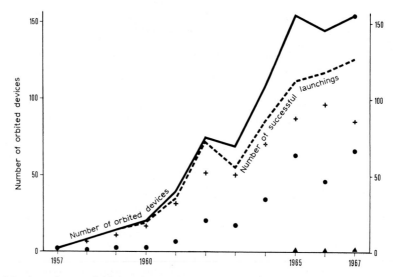

Fig. 1.   Number of successful launchings and orbited devices since 1957. During the same period, the masses placed in orbit increased very considerably: from 83.5 kg for Sputnik 1 (1957) through the 6.5 tons of Sputnik 7 (1961), up to the 126 tons of Saturn V and Apollo 4 (1967). Number of devices placed in orbit: + U.S.A.; ● U.S.S.R.; ▲ France.

1. Universal Gravitation

If we suppose perturbations absent, and supposing the mass of the earth concentrated in a single point (we shall see below to what extent this approximation is valid), then the orbit of the satellite is an *ellipse* with this point as one of its foci; the place of this ellipse is fixed with respect to the sky, cutting the celestial sphere along a specific and in principle fixed great circle.

The motion of the satellite in this orbit is then described by Kepler's laws. Moreover, the possible families of orbits are very limited, since, physically, the perigee (the point of the orbit closest to the earth, here supposed concentrated in its center) cannot be located inside the sphere constituted by the material surface of the earth (Figure 2).

Before studying these orbits let me reassure the reader; today celestial mechanics is a very difficult subject, and I plan to give only a few notions of it here – those necessary to an understanding of the methods of space astronomy – but at least I would like to establish the principal results without too much begging the question. Thus, before studying these orbits, we must calculate the potential of the gravitational forces, so as to verify the fundamental hypothesis, namely that in this sort of calculation we can take the earth to be an attracting point mass.

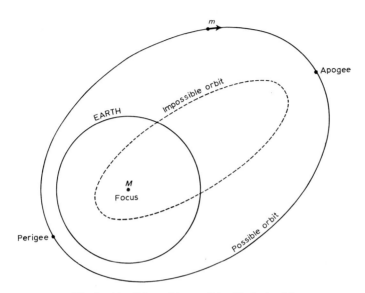

Fig. 2.   Possible and impossible elliptical orbits.

We know that at each point, the attractive forces exerted by a mass M on a mass m is a vector quantity $\mathbf{F}$, pointing from the mass m to the mass M along the straight line joining them. The magnitude of this force is $(GMm)/r^2$, and we shall write this force vector as

$$\mathbf{F} = \frac{GMm}{r^2}\,\mathbf{u},\tag{1}$$

where $\mathbf{u}$ is the unit vector pointing from m to M, and G is the universal gravitation constant.

We know (and this can be easily verified) that this force is derived from a potential, i.e. in the space surrounding the mass M, the force $\mathbf{F}$ depends only on the point chosen, and has the value

$$\mathbf{F} = -\,m\,\mathbf{grad}\,V\tag{2}$$

(or $-m\nabla V$ using the 'nabla' symbol for $\mathbf{grad}$), where V, a scalar quantity, is the gravitational potential of the mass M, whose value at each point is

$$V = -\,G\frac{M}{r}.\tag{3}$$

Newton's law of universal gravitation is expressed by either expression (1) or expression (2) together with (3). The conventional value of $GM_\oplus$, adopted in 1964 by the International Astronomical Union, is 398.603 km^3 sec^{-2} ($M_\oplus$ stands for the earth's mass).

2. Potential of a Spherical Mass

If, instead of having a single attracting mass, we must consider several masses located in different points, we find that their potentials combine by *scalar addition*, while the corresponding forces must be added *vectorially*. If we want to compute the potential at a point, or the force acting on a mass m at this point, we see that it is easier to work with the potential rather than with the more complicated vector addition.

Thus, to compute the motion of a satellite in the field of the earth, no longer supposing the latter to be a point, we must sum the contributions to the potential of the different infinitesimal portions of the attracting mass:

$$V = -G \int \frac{dm}{r}. \tag{4}$$

In order to evaluate this integral over the volume of a sphere, we shall choose an element of volume, as shown in Figure 3, obtained by decomposing the sphere into

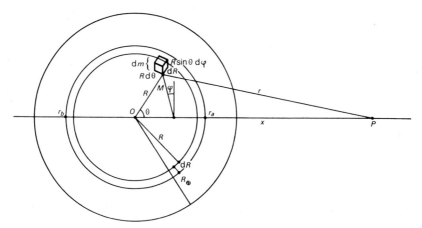

Fig. 3.   Calculation of the potential outside a spherical mass.

concentric shells. Thus the element of mass is

$$dm = \mu(R) \, R^2 \sin\theta \, d\theta \, d\varphi \, dR, \tag{5}$$

where $\mu(R)$ is the mass for unit volume. Thus at the point P, the potential of a shell of thickness dR is:

$$V = -G \int_0^\pi d\theta \int_0^{2\pi} d\varphi \mu(R) \, R^2 \frac{\sin\theta}{r} \, dR. \tag{6}$$

Using the obvious geometric relation:

$$r^2 = x^2 + R^2 - 2Rx \cos \theta \qquad (7)$$

we can write

$$r \, dr = xR \sin \theta \, d\theta \qquad (8)$$

and pursue the integration.

Equation (6) above becomes

$$V = - G \int_0^{2\pi} \int_{r_a}^{r_b} \mu(R) \, R \, \frac{r \, dr \, d\varphi}{xr} \, dR . \qquad (9)$$

In this integration, x and R are constants, so that the potential of the thin spherical shell of thickness dR and center O is:

$$V = - G \, \frac{R}{x} \, \mu(R) \, dR \int_0^{2\pi} \int_{r_a(\theta=0)}^{r_b(\theta=\pi)} dr \, d\varphi . \qquad (10)$$

The way in which we proceed with the integration depends on whether P is inside or outside the sphere of radius R. Obviously we shall suppose that our case is of 'astronomical' interest, with the point P an external point. Then

$$r_a = x - R$$
$$r_b = x + R$$

and thus

$$V = - 4\pi G \mu(R) \, \frac{R^2}{x} \, dR = - \delta m(R) \, \frac{G}{x} . \qquad (11)$$

It is just as if the mass δm of the spherical shell was concentrated in O. This is therefore equally true for all the spherical shells which make up the sphere, if we assume that *the density depends only on the distance from the center O* (which moreover is not a bad representation of the earth itself, at least *to a first approximation*).

Thus, as a first approximation, we can consider the motion of a satellite around the earth to be that of a small point mass in orbit around the earth under the attraction of another point mass $M_\oplus$, located at the center of the earth.

Unfortunately (or fortunately), given the accuracy of astronomical observations, this approximation is not perfect. Thus we are obliged to study the *perturbations* of this motion; furthermore such a study will yield valuable information of high accuracy regarding the perturbing forces; and so we see the interest here is in the dynamic study of satellite motions... We shall come back to this point later on (pp. 24–34).

However, before we go into the description of the problems really faced by the specialists on satellite motions, it will be useful to get an idea of the magnitudes involved, in a few simplified and indeed schematic cases.

3. Numerical Study of the Simple Case of Circular Orbits

From the preceding considerations, as a first approximation, we can consider the problem of two point masses as representative of the satellite problem.

Circular orbits constitute one family of possible orbits, and from them we can fix some orders of magnitude. Moreover it should be noted that most real orbits are nearly circular: since the earth's radius is about 7000 km, an orbit whose apogee is at 700 km from the ground and perigee at 350 km, has an eccentricity $e = 0.025$, which is very low. Thus it is clear that in many practical cases, the circular orbit will be a sufficiently good representation of the true orbit.

The period P is related to the semi-major axis a of the orbits (i.e. to the radius of these orbits) by Kepler's Third Law:

$$\frac{a^3}{P^2} = \frac{GM_\oplus}{4\pi^2} = \text{constant, or: } P = 2\pi(GM_\oplus)^{-1/2} a^{3/2}. \tag{12}$$

In a circular orbit, the magnitude of the velocity v (module of the velocity vector) is

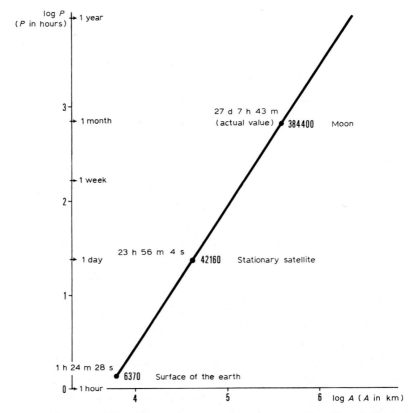

Fig. 4. Period of circular orbits around the earth. Ordinate: the logarithm of the period (in hours); abscissa: the logarithm of the distance to the center of the earth (in kilometers).

constant, and equal to

$$v_{\text{circ}} = (GM_\oplus)^{1/2} \, a^{-1/2} \, . \tag{13}$$

Figures 4 and 5 show how these quantities depend on distance. Note that the 'stationary' satellite (which appears fixed in the sky, from any fixed point on earth – provided obviously that it is launched in the same direction of rotation as the earth and not in the opposite direction, and that this rotation lies in the earth's equatorial plane) corresponds to a distance:

$a = 42\,160$ km, from the satellite to the center of the earth.

Naturally, it should be noted here that if such a satellite has an orbit inclined to the equator, it will appear to stay on a fixed meridian in the sky, as seen from the earth; but on this meridian, it will move both north and south of the equator. Furthermore,

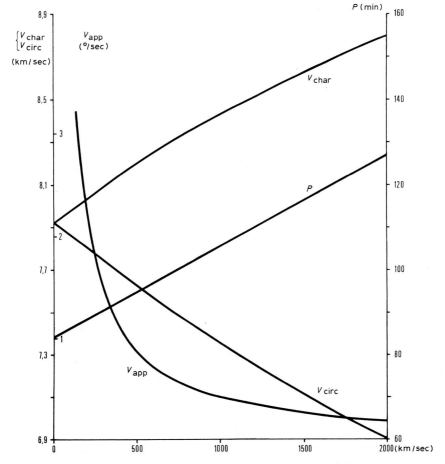

Fig. 5. Characteristics of circular orbits. Abscissa: the altitude above the earth's surface (in kilometers); ordinate: (extreme left) the characteristic and circular velocities v_{char} and v_{circ}, (left) the apparent angular velocity v_{app}, and at right, the period P.

any departure of the orbit from circularity will have as effect a variation of the speed along the trajectory, following the law of equal areas (Kepler's Second Law): then the 'stationary' satellite will appear to move along a figure-8 curve. This curve is very sensitive to the perturbations which modify the orbit, and its evolution is of great interest. For example, it leads to a determination of the ellipticity of the earth's equator.

For $a = 384400$ km (i.e. at the distance of the moon from the earth) we find a period of 27 days 12 hrs, which would correspond to the lunar period of revolution, if the moon were of zero mass; correcting for the lunar mass $M_{\mathbb{C}}$ we then have:

$$P = 2\pi (M_\oplus + M_{\mathbb{C}})^{1/2} \, a^{3/2} G^{-1/2} \tag{14}$$

and we obtain a value 27 days, 7 hr, 7 min, which is very close to the true value of 27 days, 7 hrs, 43 min.

We have mentioned that low-altitude satellites (in fact nearly all those which have had some popular celebrity) have had nearly circular orbits. Figure 5 gives the properties of strictly circular low-level satellites: period, circular velocity, maximum apparent angular velocity, as well as the quantity v_{char}, which we shall discuss later (page 45). Note that the apparent speed of passage (e.g., in degrees per second of time) decreases much faster than the circular velocity in the orbit.

4. Non-Circular Keplerian Orbits of Artificial Satellites

Although we shall not yet go into the study of perturbations, which is the principal motivation for the study of satellite motions, we must extend the preceding remarks to non-circular orbits, and define the conditions under which *satellization* occurs.

In general, artificial satellites do not have circular orbits, although as we have said, their eccentricity is often very small, close to that of a circle. Using Keplerian two-body mechanics, we shall consider for the moment that these orbits can be characterized by the period and the altitude of the perigee alone – or by the altitude of the perigee and that of the apogee. These orbits may correspond to an injection (after some flight of the launch vehicle: obviously the only practical case in general) from some point of the orbit: the magnitude and the direction* of the *initial velocity* will determine the orbit completely, and once the data of this initial velocity, plus the point of injection, are chosen, we have enough parameters for the calculation (Figure 6).

We shall restate here a number of very classical results of Keplerian mechanics.

First of all, for a launch point on the earth, we shall note that unless the launch velocity is both strictly horizontal and equal to or greater than the circular velocity, the orbit will intersect the earth's surface at another point (inasmuch as an ellipse is symmetrical about its major axis) and the trajectory will come to a sudden stop; thus the launch point should be elevated; further on we shall see how the initial point of a Keplerian orbit may be attained, by way of an intermediate, 'transfer' orbit, and using non-gravitational forces.

---

* Note that the symbol V_0 (used further on for the angle of the radius vector OP_0 to the velocity) should not be confused with the symbol V (used above for the potential).

Clearly if the initial velocity is very large, the device will be able to leave the earth and its gravitational field and go on to 'infinity' (i.e. in this case in the sun's gravitational field). On the contrary, if the initial velocity is very low, the device will fall back down, and will be stopped by the earth's surface during its elliptical fall toward a 'pericenter' inside the earth (which clearly can no longer be correctly called a perigee!).

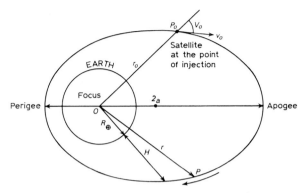

Fig. 6. Parameters determining the orbit.

Moreover, we shall note that this case (the usual case of ballistics) is generally treated not as a central force problem as here, but as a problem involving a uniform force field, the 'gravitational field' in the neighborhood of the launch point. In that case one finds that the trajectory is a parabola, osculatory to the ellipse of the Keplerian central force problem: the approximation is a good one, to the same accuracy with which the earth's surface can be represented by a plane.

When the velocity is insufficient ($V_0 = \pi/2$, $v_0 < v_{\text{circ}}$) or non-horizontal ($V_0 < \pi/2$), we are dealing with ballistics, such as that concerned with missiles. The projectile will fall back to earth, and we may want to know its range. Using Kepler's laws, the computation is simple. Indeed we have (anticipating Equation (27), which we shall see below, and which is well known):

$$\varphi_0 = \text{arc cos} \frac{1 - \alpha^2 \sin^2 V_0}{\sqrt{[1 - (2 - \alpha^2) \alpha^2 \cos^2 V_0]}},\tag{15}$$

where α is defined by:

$$\alpha = v_0/v_{\text{circ}}.\tag{16}$$

This range takes its maximum value for:

$$\sin^2 V_0 = 1/(2 - \alpha^2)\tag{17}$$

and this maximum value is equal to:

$$\varphi_{0,\,\text{max}} = \text{arc sin} \frac{\alpha^2}{2 - \alpha^2}.\tag{18}$$

Let us come back to complete Keplerian orbits. The type of orbit is determined by the *energy constant h.*

$$h = v_0^2 - 2GM_\oplus/r_0. \tag{19}$$

If $h \geqslant 0$ (*hyperbolic* or *parabolic* orbits), the projectile does not describe a closed orbit. The direction of its velocity at infinity depends on the orientation of the initial velocity: but the magnitude of the velocity at infinity depends only on the magnitude of the initial velocity. For a parabolic orbit it goes to zero; for a hyperbolic orbit it is just equal to the square root of the energy constant:

$$v_\infty = v_0 \left(1 - 2G\frac{M_\oplus}{v_0 r_0}\right)^{1/2}. \tag{20}$$

When h is less than zero, the orbits are *elliptic*. The orbit then has the following characteristics, which are determined by the initial conditions V_0, r_0, v_0 (these relations are well known, and for their proof we refer the interested reader to a treatise on conic sections or to courses in classical mechanics).

The semi-major axis of the ellipse is given by:

$$a = -\frac{GM_\oplus}{h} = \frac{GM_\oplus}{2GM_\oplus/r_0 - v_0^2}. \tag{21}$$

The period is:

$$P = 2\pi a^{3/2}(GM_\oplus)^{-1/2} = 2\pi GM_\oplus\left(\frac{2GM_\oplus}{r_0} - v_0^2\right)^{-3/2}. \tag{22}$$

The eccentricity e is obtained from:

$$1 - e^2 = \frac{C^2}{GM_\oplus a}, \tag{23}$$

where C is the area constant which appears in Kepler's Second Law.

$$C = \frac{r^2\,d\varphi}{dt} = \text{constant} = r_0 v_0 \sin V_0. \tag{24}$$

Thus we have:

$$1 - e^2 = \frac{r_0^2 v_0^2 \sin^2 V_0}{(GM_\oplus)^2}\left(\frac{2GM_\oplus}{r_0} - v_0^2\right). \tag{25}$$

The ends of the major axis are found from:

$$r_{\min} = a(1 - e), \quad r_{\max} = a(1 + e), \tag{26}$$

and these are the *perigee* and the *apogee*.

The motion of the satellite on such an orbit must be determined.

The location of the satellite on the ellipse is fixed most conveniently by the polar coordinates r and φ (angle called the 'true anomaly'), where the relation

$$r = a(1 - e^2)/(1 + e \cos \varphi) \tag{27}$$

fixes the ellipse of the orbit. These two quantities vary as a function of time, and the magnitude v of the velocity is given by:

$$v = (GM_\oplus)^{1/2} \left[\frac{2}{r} - \frac{1}{a} \right]. \tag{28}$$

The elliptic orbit must satisfy a second condition (the first was $h < 0$):

$$r_{\min} = a(1 - e) > R_\oplus, \tag{29}$$

which assures that the orbit remains above the earth's surface.

Considering $r_0 = R_\oplus + H_0$ as given (where H_0 is the altitude of the satellite at injection), and allowing only v_0 and V_0 to vary, we can describe possible orbits quite simply as a two-parameter family.

If $r_0 = R_\oplus$, the first condition to be satisfied takes the form

$$v_0 < 2GM_\oplus/R_\oplus = 11.2 \text{ km/sec} = v_{p, 0},$$

whatever V_0 may be. We can have an elliptic orbit if and only if v_0 is less than the 'parabolic' velocity $v_{p, 0}$.

In the following discussion, we shall use a rather different notation, *normalizing* in particular the unit of velocity. We shall write y for the square of the normalized velocity: $y = (v_0/v_{p, 0})^2$. We shall denote the non-dimensional ratio $H_0/R_\oplus$ by ε, and $\sin^2 V_0$ by x.

The basic Equation (25) above becomes:

$$z = 1 - e^2 = 4xy(1 + \varepsilon)^2 \left(\frac{1}{1 + \varepsilon} - y \right)$$
$$= 4xy(1 + \varepsilon)[1 - (1 + \varepsilon) y]. \tag{30}$$

It is immediately evident that the two quantities which determine the eccentricity of the orbit are x (direction of firing) on the one hand, and the product $(1 + \varepsilon) y = \eta$ on the other.

From this relation we can see that at an arbitrary height H_0 the parabolic velocity $v_{p, H}$ corresponds to $\eta = 1$, and is given by:

$$v_{p, H} = v_{p, 0}/\sqrt{1 + H_0/R_\oplus}. \tag{31}$$

The quantity $z/x = 4\eta(1 - \eta)$ varies, as shown in Figure 7, and for $\eta = \frac{1}{2}$, z/x is unity. This value corresponds to a velocity

$$v_0 = \left(\frac{GM_\oplus}{R_\oplus} \right)^{1/2} \left(1 + \frac{H_0}{R_\oplus} \right)^{-1/2} = v_{\text{circ}}, \tag{32}$$

and this is the value which appears in Figure 5 for $H_0 = 0$.

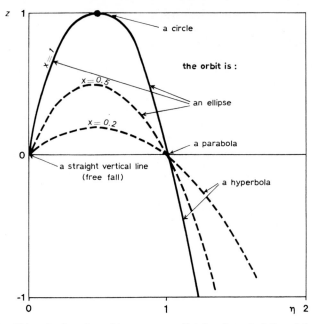

Fig. 7.   Using the function $z(\eta)$ one can predict the characteristics of the orbit.

Note that if $x=1$ (horizontal injection), this value does indeed correspond to a circular orbit. When v_0 takes this value, the type of orbit changes: for $v_0 > v_{circ}$, the center of the earth is the nearer of the two foci to the injection point, which is then the perigee of the orbit; when $v_0 = v_{circ}$ the two foci coincide with the center of the earth; for $v_0 < v_{circ}$ the center of the earth is the more distant of the foci, and the injection point is the apogee (Figure 8).

If $x<1$ (inclined injection), the orbit can never be circular: as v_0 decreases the axis of the orbital ellipse rotates and the eccentricity goes through a minimum without reaching zero (Figure 9).

Let us return now to the second condition for an orbit:

$$r_{min} = a(1 - e) > R_\oplus .\tag{33}$$

For $x=1$ we can write quite simply (from Figure 6):

$$2a > H_0 + 2R_\oplus ,\tag{34}$$

where, using (21) and the definitions of ε and η,

$$\frac{a}{R_\oplus} = \frac{1}{2}\frac{1 + \varepsilon}{1 - \eta} .\tag{35}$$

Then we have

$$\frac{1 + \varepsilon}{1 - \eta} > \varepsilon + 2\tag{36}$$

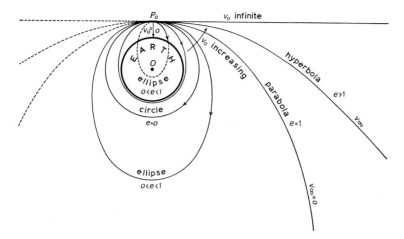

Fig. 8.   Families of orbits corresponding to a horizontal launching or injection.

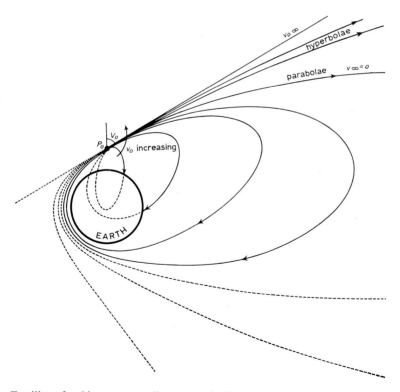

Fig. 9.   Families of orbits corresponding to an inclined injection. Note that a circular orbit is impossible.

or

$$\eta > \frac{1}{2 + \varepsilon} = \tfrac{1}{2} - \tfrac{1}{4}\varepsilon \quad (\text{if } \varepsilon \text{ is small}). \tag{37}$$

Thus the orbit will intersect the earth when v_0 is smaller than the value:

$$v_{\text{fall}} = \left(\frac{2GM_\oplus}{R_\oplus}\right)^{1/2} [(\tfrac{1}{2} - \tfrac{1}{4}\varepsilon)(1 + \varepsilon)^{-1}]^{1/2}. \tag{38}$$

As one might have guessed intuitively, this value is slightly smaller than the circular velocity.

If on the contrary, x is different from 1, the calculation is somewhat more complicated but still feasible. We shall leave it for the edification of the interested reader. The condition above is no longer valid since injection no longer is taking place at the apogee end of the major axis. In any case it is easy to see that the injection velocity must be directed outside of a cone tangent to the terrestrial globe with its apex at the injection point (Figure 10).

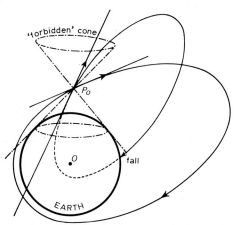

Fig. 10. Definition of the conditions of injection leading to possible orbits.

Note that in all of these problems, it is the square of the sine of the angle of inclination of the initial velocity to the vertical that enters the calculation. In other words, for each value V_0 there will be a value $\pi - V_0$ leading to an identical orbit.

In the absence of perturbations, the problem of satellite orbits reduces to the above problems. Specialists in celestial mechanics would lose interest quickly, were it not for the present accuracy of observations, which makes possible a study of *perturbations* of these orbits, and also for the problem of *transfer orbits*, which by the introduction of forces at a given point enable one to pass from one gravitational orbit to another.

5. Perturbations of Keplerian Orbits

There are many such perturbations, and the most important are those due to the asphericity of the earth.

A. PERTURBATIONS DUE TO THE ASPHERICITY OF THE EARTH

A supplementary attractive force must be added to the central force, and this perturbing force is of variable direction. Schematically this force can be described as being essentially caused by the equatorial bulge of the terrestrial globe. Clearly, the study of this sort of perturbation of artificial satellite motions is of great importance in geodesy. It should be emphasized that for artificial earth satellites this cause of perturbations dominates by far all the other gravitational perturbations we shall mention, because of the very small distance of the satellite from the earth.

To begin the calculation of perturbed orbits we need the potential V of the earth, given by Equation (4). If we assume R/x sufficiently small, we can write Equation (7) as:

$$\frac{1}{r} = \frac{1}{x}\left(1 + \frac{R}{x}f_1(\theta) + \frac{R^2}{x^2}f_2(\theta) + \cdots\right). \tag{39}$$

The functions $f_i(\theta)$ are the Legendre polynomials of the variable $\cos\theta$:

$$f_i(\theta) = P_i(\cos\theta), \tag{40}$$

where

$$P_n(x) = \frac{(2n)!}{2^n(n!)^2}\left\{x^n - \frac{n(n-1)}{2(2n-1)}x^{n-2}\right. \\ \left. + \frac{n(n-1)(n-2)(n-3)}{2\cdot4\cdot(2n-1)(2n-3)}x^{n-4} + \cdots\right\}. \tag{41}$$

We can then easily transform Equation (4):

$$V = -G\int\frac{dm}{r} = -G\int\frac{dm}{x}\left[1 + \frac{R}{x}f_1(\theta) + \cdots\right] \\ = V_0 + (V_1 + V_2 + \cdots) = V_0 + \mathscr{R}. \tag{42}$$

The zero-order term $V_0 = -GM/x$ is just the term we found earlier.

Going back to Figure 3, if we consider the mass dm at point M (coordinates ξ, η, ζ) and the point P (coordinates α, β, γ) at which we are computing the gravitational potential, we find

$$\cos\theta = \frac{\alpha\xi + \beta\eta + \gamma\zeta}{Rx} \tag{43}$$

and

$$V_1 = -\frac{G}{x^3}\left[\alpha\int\xi\,dm + \beta\int\eta\,dm + \gamma\int\zeta\,dm\right]. \tag{44}$$

The integrals $\int\xi\,dm$ give the coordinates of the center of gravity of the masses dm (and are equal to zero if the origin is taken at the center of gravity).

We can compute V_2, V_3, V_4 in the same way. The expression for V_2 includes the moments and products of inertia of the body with respect to the coordinate axes. Terms V_3 and V_4 include higher-order moments of inertia.

For a spherically symmetrical body, made up of homogeneous shells (the case treated on page 14), all these terms cancel and only the term V_0 remains.

In current research on the earth's potential, it is represented as a function of latitude φ and longitude L:

$$
V = \frac{GM_\oplus}{r} \left\{ 1 - \sum_{n=2}^{\infty} \left[\mathscr{J}_n \left(\frac{R_\oplus}{r} \right)^n P_n(\sin \varphi) \right. \right.
$$
$$
\left. \left. + \sum_{m=1}^{n} \left(\frac{R_\oplus}{r} \right)^n p_n^m(\sin \varphi)(C_{nm} \cos mL + S_{nm} \sin mL) \right] \right\}. \tag{45}
$$

In this expression the P_n are the Legendre polynomials and p_n^m the associated Legendre polynomials:

$$
p_n^m(x) = \sqrt{\frac{(n+m)!}{2(2n+1)(n-m)!}} (1-x)^{m/2} \frac{d^m P_n(x)}{dx^m}. \tag{46}
$$

The $\mathscr{J}_n$ are the 'zonal harmonics':

$$
\begin{aligned}
\mathscr{J}_2 &= 1082.64 \times 10^{-6} \\
\mathscr{J}_3 &= -2.55 \times 10^{-6} \\
\mathscr{J}_4 &= -1.65 \times 10^{-6} \\
\mathscr{J}_5 &= -0.21 \times 10^{-6} \quad \text{(according to Kozai, 1964)}.
\end{aligned}
$$

The C_{nm} and the S_{nm} are 'tesseral harmonics':

$$
\begin{aligned}
C_{22} &= 2.45 \times 10^{-6} \\
S_{22} &= -1.52 \times 10^{-6} \quad \text{(Anderle, 1965)}.
\end{aligned}
$$

Their meaning is as follows:

$\mathscr{J}_2$ measures the equatorial bulge of the earth;

$\mathscr{J}_3$ measures the asymmetry between the northern and southern hemispheres (pear-shaped component of the earth's figure); C_{22} and S_{22} measure the ellipsoidal shape of the earth's equator. The high accuracy presently available from the analysis of satellite observations is made evident by the fact that more than 200 coefficients C_{nm} and S_{nm} are known!

These coefficients must be determined if one wishes to describe (and predict) the trajectories within an accuracy of about 20 m. Data obtained by lasers will certainly enable us to reduce this error margin still further, and to determine still more coefficients $\mathscr{J}_n$, C_{nm} and S_{nm}.

B. OTHER GRAVITATIONAL PERTURBATIONS

Lunar perturbations are significant for artificial satellites, and as with tidal forces, they are stronger than the solar attraction. *A fortiori*, when earth-moon probes are involved, the problem can no longer be considered one of perturbations, but rather as a 'three-body problem', in which only one body is of negligible mass.

Solar perturbations remain very important (and indeed fundamental in the case of the motion of a natural satellite, such as the moon! about its planet). In the case of interplanetary probes (say between the earth and another planet) it is the sun which is the principal attractive mass, while on the contrary it is the earth, or e.g. Mars, that perturbs the vessel's motion in the solar gravitational field (see below, p. 47).

As for perturbations due to other planets or artificial satellites, these are negligible for artificial earth satellites, except for the case of a very close passage of another satellite. In such a case we are in fact dealing with a 'collision'; i.e. there is an exchange of energy and passage from one Keplerian orbit to another, the transition taking place during a time interval and along a segment of the trajectory very short compared to the period and the circumference of the orbit, respectively. It may well be that the second orbit (after the 'collision') does not belong to the family of permitted orbits: the fall and disappearance of the satellite are then inevitable. However, the number of satellites in circulation is still small enough for this possibility to remain quite unlikely.

C. NON-GRAVITATIONAL PERTURBATIONS

Besides the perturbations due to the earth's asphericity, some of these play an equally significant role.

Artificial satellites of the earth move in an atmosphere which remains sufficiently dense for *frictional* forces to be non-negligible. Since these forces depend on the satellite's altitude, they are variable if the orbit is not strictly circular.

The effect of solar *radiation pressure* is also very important, the more so when the satellite has a large surface exposed to the sun's rays (as is the case, e.g., for satellites like Echo). Obviously radiation pressure effects are discontinuous at the moment the satellite leaves the sunlit zone for the region of the earth's shadow (or vice versa). *Dynamic pressure* effects due to the solar 'wind', and to rapid dense clouds of corpuscles linked to solar flares, are certainly far from negligible, but remain poorly understood. There has been even less study of the effects of *electrostatic* or *electromagnetic* forces.

D. ARTIFICIAL PERTURBATIONS

Finally, we must underline the particularly high interest of artificial perturbations, generated by radio command, by automatic programming within the device, or by the maneuvers of an astronaut. These perturbations are produced by small rockets, and make possible the correction of the orbit or trajectory in mid-course. Thus one can compensate for errors in firing; also one can obtain trajectories which are not Newtonian orbits. We shall return to this sort of problem in more detail; clearly this is what astronautics is really about. However, to the extent that that control of these devices can be permanent, as say that of an automobile, their dynamics can hardly be used to study the natural forces influencing them. This is no longer experimental celestial mechanics!... This is why we shall omit this type of perturbation here, but we shall consider it later on (pp. 46–51).

6. Detailed Examination of Gravitational Perturbations

The study of gravitational perturbations is basically that of the three-body problem (satellite, earth – assumed concentrated in its center – and perturbing masses). It is well known that this problem can be very complicated, and clearly we can only deal with certain necessarily limited aspects here. Let us first examine a very simple case.

A. DETERMINATION OF THE NEUTRAL POINT

A first way by which the problem can be (severely!) simplified is to limit it to a one-dimensional problem, in which (e.g.) we determine the 'neutral' point between the earth and the moon, where the attractive forces exerted by the two masses are equal; then we can study certain properties of this unique and singular point.

Fig. 11.

Let $M_\oplus$ and $M_\mathbb{C}$ be the masses of the earth and the moon, respectively, located at O and O' (Figure 11). The equation of motion will then be:

$$v \frac{dv}{dr} = -\frac{GM_\oplus}{r^2} \pm \frac{GM_\mathbb{C}}{(a \mp r)^2}.$$ (47)

In this equation, the lower sign applies to a body located at P on the earth–moon line, but outside the earth-moon segment. Clearly we can integrate this and write

$$v^2 = v_0^2 + 2GM_\oplus \left(\frac{1}{r} - \frac{1}{r_0} \right) \pm 2GM_\mathbb{C} \left[\frac{1}{a \mp r} - \frac{1}{a \mp r_0} \right].$$ (48)

In order to compare these results with those for the earth-satellite system, we must introduce into the equations the parabolic velocity at a distance r_0 from the center of the earth O or at a distance b from the moon's center O' (computed assuming both earth and moon isolated in space). Then we can write (see above, p. 20):

$$v_p^2(r_0) = 2GM_\oplus/r_0$$ (49)

and

$$v_p^2(b) = 2GM_\mathbb{C}/b,$$ (50)

and consequently

$$v^2 = v_0^2 + v_p^2(r_0) \left[\frac{1}{r/r_0} - 1 \right] \pm v_p^2(b) \left[\frac{1}{a/b \mp r/b} - \frac{1}{a/b \mp r_0/b} \right].$$ (51)

Position of the Neutral Point. – The equality of the two forces of attraction, of the moon and of the earth, may be written:

$$\frac{GM_\oplus}{r_n^2} = \frac{GM_\mathbb{C}}{(a - r_n)^2},$$ (52)

so that we have:

$$r_n = \frac{a}{\sqrt{1 + M_\oplus/M_\mathbb{C}}},$$ (53)

using the data in the Appendix, Table VII, we find

$$r_n = 51.2 \, R_\oplus.$$ (54)

Note that $a = 60.258 \, R_\oplus$.

What is the Initial Velocity needed to reach the Moon from the Earth along a Straight-Line Trajectory? – Let us set $r = r_n$ and $v = 0$ in Equation (48); the resulting velocity will be the minimum initial velocity for the trip to the moon:

$$v_0 = 11.05 \text{ km sec}^{-1},$$ (55)

which is slightly lower than the parabolic escape velocity ($v = 11.18$ km sec^{-1}). Clearly however one cannot launch a vessel from the earth to the moon without taking into account the motions and rotations of the two bodies: therefore the above approximation is very poor.

What is the Initial Velocity needed to escape from the Sun's Gravitational Field, Starting from the Earth? – In the preceding relations, we must replace the earth by the sun and the moon by the earth. From the data concerning the sun and the earth (Appendix, Tables VI and VII), we obtain the value 43.7 km sec^{-1}. In fact, the earth moves in its orbit around the sun, and so a satellite of the earth already possesses a certain motion around the sun. Thus the supplementary velocity needed for escape is lower than the value computed above. We shall discuss later (pp. 42ff.) more of the details of the problems of space probes which leave the zone where the earth's attraction is predominant.

B. GENERAL CALCULATION OF GRAVITATIONAL PERTURBATIONS

The general calculation of gravitational perturbations is obviously much more complex than the extremely simple case we have just discussed.

In principle, Lagrange's equations permit a completely general formulation. However, in the case of artificial satellites, the perturbations can still be taken to be

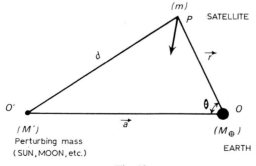

Fig. 12.

sufficiently small, and so can be treated independently. In this case, if M' is a perturbing mass at a distance a from the earth's center (Figure 12), the equation of motion is:

$$\frac{d^2\mathbf{r}}{dt^2} = - G\left(\frac{M_\oplus + m}{r^3}\right)\mathbf{r} + GM'\left(\frac{\mathbf{d}}{d^3} - \frac{\mathbf{a}}{a^3}\right). \tag{56}$$

Introducing a 'perturbing function' $\mathscr{R}$ we can also write:

$$\frac{d^2\mathbf{r}}{dt^2} = - G\left(\frac{M_\oplus + m}{r^3}\right)\mathbf{r} + \mathbf{grad}\,\mathscr{R}, \tag{57}$$

where

$$\mathscr{R} = GM'\left(\frac{1}{d} - \frac{\mathbf{a}\cdot\mathbf{r}}{r^3}\right). \tag{58}$$

The perturbing function $\mathscr{R}$ was written earlier in a more general form, in which M' was not taken to be a point mass. However, in Equation (42) $\mathscr{R}$ was taken to be a function of the latitude and longitude of the point on the earth at which the satellite appeared in the zenith (subsatellite point); essentially the perturbing masses were assumed to be attached to the earth.

It is better to write $\mathscr{R}$ as a function of the six parameters of the Keplerian ellipse osculatory to the trajectory (semi-major axis a, eccentricity e, inclination i, right ascension of the ascending mode Ω, argument of the perigee ω, mean anomaly $\mathfrak{m}$: see Figure 16).

The manipulation of Lagrange's equations leads to the calculation of the evolution of the parameters of the osculatory ellipse. Solution is possible only by using approximation methods. We can go no further here, and refer the reader to J. Kovalevsky's book.* Still it should be clear that with this generalized theory one can introduce a function $\mathscr{R}$ which is time-dependent as well as variable in space, and so calculate the perturbations due to the moon and the sun.

C. TIME-DEPENDENT TERMS

The final solution takes the form of a series expansion which contains four sorts of time-dependent terms for each orbital parameter.

(1) *Secular Terms.* – These terms are linear in time. The motion of the perigee (rotation of the 'line of apsides' – the major axis – in its plane) and that of the ascending node (rotation of the plane of the orbit about the earth's axis) appear in such terms. These two motions are essential properties of the motion of an artificial satellite, whereas the others are simply oscillations about some constant mean term, or some term linear in time.

Since these terms are related to the function $\mathscr{R}$ corresponding to the gravitational potential of the (aspherical) earth, they can be calculated if we know the potential. The expressions mentioned above allow us to compute, in degrees per day, the

* J. Kovalevsky, *Introduction to Celestial Mechanics*, Springer, New York; D. Reidel, Dordrecht, 1969.

motion of the perigee:

$$\frac{\partial \omega}{\partial t} = 4.98 \left(\frac{R_\oplus}{a}\right)^{3/2} (1 - e^2)^{-2} (5 \cos^2 i - 1) \tag{59}$$

as well as the linear component of the motion of the ascending node:

$$\frac{\partial \Omega}{\partial t} = 9.97 \left(\frac{R_\oplus}{a}\right)^{3/2} (1 - e^2)^{-2} \cos i \tag{60}$$

as functions of the orbital parameters (see Figure 16). In practice, these terms are of the order of magnitude of a few degrees per day. It can be seen that two values of the inclination: $i = 90°$, and $i = 63°4$ ($\cos i = \sqrt{\frac{1}{5}}$), play a special role. For a polar orbit ($i = 90°$), $\partial \Omega / \partial t = 0$; thus when $\cos i$ passes through the value $\cos i = 0$, the direction of rotation of the plane of the orbit changes.

For $i \neq 63°4$ ('critical' inclination) we have libration of the perigee: $i < i_{crit}$, the perigee moves in the same direction as the satellite; for $i > i_{crit}$, in the opposite direction.

(2) *Periodic Terms of Long Period.* – These terms are comparable to those of the lunar libration. Their period is related to the motion of the perigee in the plane of the orbit, and is of order of magnitude two to three months. The most striking terms are due to the presence of the $\mathscr{J}_3$ terms (asymmetry of the two terrestrial hemispheres) in the expansion of the potential. In particular large variations of eccentricity are observed. The altitude of the perigee oscillates with an amplitude of a few tens of kilometers.

(3) *Terms whose Period is Close to a Day.* – These are due to the longitude-dependent part of the potential and are related to the earth's rotation. They are due, e.g., to the ellipticity of the earth's equator.

(4) Finally we have the *Solar and Lunar Terms.* – These are negligible for low-level satellites, and only become of importance for very distant satellites (highly eccentric orbits), as well as naturally for space probes.

7. Non-Gravitational Perturbing Forces

Forces of non-gravitational origin (frictional forces, forces due to radiation pressure) play an essential role. It is practically impossible to devise an analytical theory for them. The equations of motion can be solved only by suitable numerical methods, so as to evaluate the perturbing factors by analysis of the observations.

The equations to be solved are of the form

$$\frac{d^2 \mathbf{r}}{dt^2} = - G \left(\frac{M_\oplus + m}{r^3}\right) \mathbf{r} + \mathbf{f} \left(\frac{d\mathbf{r}}{dt}, \mathbf{r}, t\right). \tag{61}$$

If we are dealing with a frictional force the perturbing function is of the form

$$F_{frict} = - kS\rho \left|\frac{d\mathbf{r}}{dt}\right|^2, \tag{62}$$

where $|\mathbf{dr}/dt|$ is the magnitude of the velocity of the satellite relative to the atmosphere; ρ is the density of the atmosphere, S the 'collision' cross-section of the satellite, and k depends on the shape of the satellite and is of order of magnitude unity. If we are dealing with another type of force, the expression of the perturbing function can become extremely complicated, and can even become discontinuous, as for example, at the transition from sunlit to shadow zones.

Radiation pressure exerts a force F_{rad} on the satellite, where:

$$F_{rad} = (2\Phi/c)\,S \tag{63}$$

in the case of a perfect reflector, or

$$F_{rad} = (\Phi/c)\,S \tag{64}$$

in the case of a blackened perfectly absorbing satellite. If the satellite is spherical, of radius σ_0, $S = \pi\sigma_0^2$. We write Φ for the solar radiative flux, in ergs per second and per cm^2. We can estimate the magnitudes of these forces; in general they are smaller than the solar gravitational attraction. This force is

$$F_{sun} = G\,\frac{M_\odot m}{D_\oplus^2}, \tag{65}$$

where $M_\odot$ is the solar mass and $D_\oplus$ the distance from the sun to the earth. Then

$$\frac{F_{rad}}{F_{sun}} = \frac{\Phi D_\oplus^2 S}{cGM_\odot m} \sim \frac{K}{\rho\sigma_0} \sim \frac{6 \times 10^{-5}}{\rho\sigma_0}. \tag{66}$$

For a satellite of low density, this ratio becomes significant: e.g., this is the case of the Echo satellites, which consist of a hollow metallic sphere of large diameter; the radius is about 50 m, the mass about 40 kg; therefore we have

$$\rho \sim 10^{-7}, \quad \sigma_0 \sim 5 \times 10^3, \tag{67}$$

and so

$$F_{rad} \sim F_{sun}/10. \tag{68}$$

In the case of a satellite like the French satellite D1, m is about 50 kg, the diameter is about 50 cm, and so:

$$\rho \sim 10^{-1}, \quad F_{rad} \sim 10^{-5}F_{sun}. \tag{69}$$

Even in the case of Echo, a force of this magnitude is difficult to detect, but after numerous orbits of Echo its cumulative effect becomes detectable, and the computations must take it into account.

Once the perturbing forces are known, the variations of the orbital parameters can be determined by solving the equations of motion. At a given moment, using a step-by-step integration method in time one can compute the right-hand side of the equation of motion (61). The acceleration vector is then known, and taking the right-hand side constant during the following time step one continues the integration. The accuracy of the method depends essentially on the size of the integration step, i.e. the time interval separating the points at which the right-hand side is evaluated. Obviously this interval must be very small compared to the period of the satellite.

Naturally the fact that we do not know what the perturbing forces are raises difficulties: we must assume them given; from this point on comparison of theory and observation enables us to evaluate any type of perturbing term, by successive approximations.

To take an example, let us look at how one treats the problem of the evolution of an elliptical orbit under the influence of frictional forces. First of all we shall note that since the frictional force works against motion, it will reduce the energy of the satellite (transferring it to the molecules of the atmosphere). If we assume that this effect acts only at the perigee, because of the very rapid decrease of atmospheric density with altitude, the potential energy will stay constant at each perigee passage, and thus energy will be transferred to the atmosphere at the expense of kinetic energy. Thus the satellite will behave as though its succeeding orbits were Keplerian orbits corresponding to different initial velocities at the point of injection, each one being lower than the previous one (Figure 13). The energy at perigee is known (see above,

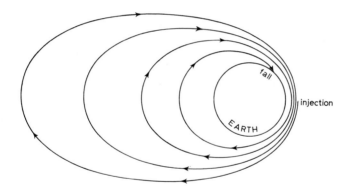

Fig. 13. Braking effect of friction in the atmospheric layers (schematic).

Equation (27)) and may be written:

$$E = GM_{\oplus}m\left[\frac{1}{r_{\text{per}}} - \frac{1}{r_{\text{per}} + r_{\text{apo}}}\right] = GM_{\oplus}\frac{mr_{\text{apo}}}{r_{\text{per}}(r_{\text{per}} + r_{\text{apo}})}. \tag{70}$$

From one revolution to another the apogee approaches the earth and E decreases. Thus from one orbit to the next, the loss of kinetic energy is:

$$\Delta E = \Delta r_{\text{apo}}m\frac{GM_{\oplus}}{r_{\text{per}}^2} \tag{71}$$

since, to second-order accuracy, we may write:

$$r_{\text{per}} \sim r_{\text{apo}} \sim R_{\oplus}.$$

However, during a single orbit, the frictional energy transferred to the medium is given by the integral:

$$\Delta E_{\text{frict}} = kS \int \rho v^2 \, ds \tag{72}$$

evaluated along the orbit.

These two expressions must be equal, and their equality yields a determination of the variation Δr_{apo} of the height of apogee. Let us look at the numbers for a typical satellite. We shall take a perigee at 200 km, an apogee at 600 km; the mass is 50 kg, the radius 50 cm. How much will the apogee be lowered during a single revolution? Given that

$$\int \rho v^2 \, ds \approx \overline{\rho} \, \overline{v^2} \int ds, \quad \overline{\rho} = 10^{-14} \quad \text{and} \quad \overline{v^2} = v_{\text{per}}^2$$

the above equation yields:

$$\Delta r_{\text{apo}} = \frac{kS}{m} f(\text{orbit}) = 80 \text{ m} \quad (k = 1). \tag{73}$$

We can estimate the satellite's lifetime. At the rate we have found, 5000 revolutions are needed to circularize the orbit: once this is done, atmospheric friction acts very strongly at all points of the orbit, and we can assume that the satellite will disappear soon afterwards. Therefore the lifetime is of the order of 3 years. In the case of the satellite D1 (perigee at 500 km) the lowering of the apogee is about 50 cm per orbit.

Observations enable us to correct the initial data on the density of the atmosphere. More physical methods also make possible a determination of the magnitudes of the various perturbations which affect the satellite's motion. In this way we can study correlations between solar phenomena and satellite motions. We shall come back to this later. At this point we must describe the astrometric procedures employed in the study of satellites.

8. Satellite Observation and Analysis of the Measurements

As we have shown just now, it is clear that observation of the orbits of artificial celestial bodies makes it possible to detect the perturbations influencing them. The study of these perturbations is one of the essential goals of the observation of artificial satellites. We shall give a rapid survey of the methods of observation and their limitations; these methods are related to the most classical astronomical methods.

A. THE DETERMINATION OF ANGULAR COORDINATES

A first remark is necessary regarding the accuracy of positional determinations and of the corresponding timings.

Low-level satellites have an apparent velocity of order of magnitude 40' per second in the sky (Figure 5). Therefore an average accuracy of 10^{-3} sec corresponds to a positional accuracy of 0".04. Thus it is quite useless to do better than 1" in accuracy in measuring the positions of such satellites. On the contrary, for objects which travel far from earth (Luniks, Pioneers, etc.) the limit of accuracy of the measurements is fixed by the positional measurements. Figure 14, which uses Dommanget's data, shows the limits of accuracy for measurements of position and of time.

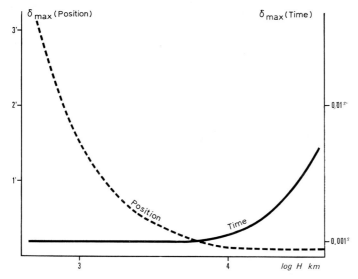

Fig. 14.   Maximum accuracy of measurements of position and of time (according to Dommanget). Abscissa: logarithm of the height (in kilometers) of the satellite above the ground.

Moreover, it is difficult to reach these limits. In any case one cannot do better: the limits are astronomical (scintillation, refraction), as well as instrumental (diffraction). Improvement will be possible once extra-terrestrial astrometric observatories are established: we haven't got there yet. The observational methods are at present still highly classical and use instruments similar to those of positional astronomy (both for visual and photographic studies), in particular very close to those used for finding comets and asteroids. The major difference arises from the fact that equatorial mountings and drives are no better than for example azimuthal ones, when it comes to practical cases. The apparent trajectory is in fact, except perhaps for very distant objects, hardly influenced by the diurnal motion of the celestial sphere.

1. *Visual Observations*   **1701267**

Here positions must be measured with respect to field stars. Some theodolites can be built for this purpose, with rapid reading of their direction possible. Once the object one is looking for is found in the field, it is centered on the cross-hairs of the reticule, and followed. When it is well centered, the instrument is fixed, the time accurately recorded (at the exact instant the satellite passes at the cross-hairs), and one reads the position. During a single satellite passage, up to 50 or 100 fixes – several per minute – can be obtained. Naturally, well-trained personnel, and as much automation as possible of the readings, using e.g. tape recorders, are necessary. Such measurements may easily have as high an accuracy as $0°\!.1$ and $0°\!.1$. With the usual light instruments, 8th or 9th magnitude objects can be reached.

2. *Photographic Observations*

Naturally, photographic observation can be of higher accuracy. Generally the objective is of such large aperture (since it must first of all be 'fast') that the diameter of the diffraction ring is smaller than the photographic grain, rather than the contrary. Thus the plate grain limits the accuracy of photographic determinations of position. With the excellent Baker-Nunn camera whose aperture is 50 cm, with an objective of diameter 78 cm and a focal length of 50 cm, positions accurate to $2''$ can be obtained.

Two different methods of observation may be used. The instrument can be adjusted so as to follow the stellar field: the photograph then shows the stellar field just as on an astronomical photograph, with the satellite's track appearing as a line across the picture. This line is interrupted by a shutter which operates at regular and well defined time intervals. By measuring the positions of the breaks in the photographed track we can determine the trajectory. The instantaneous exposure time at each point of the track clearly depends on the transit speed and the brightness of the satellite. Distant satellites can be photographed down to the 10th magnitude; closer (and therefore more rapid) satellites only to the 6th magnitude. However, their nearness compensates for this limitation, since (all other things being equal) they are brighter than more distant satellites, and fortunately this is the more important effect.

This limitation still remains. It can be overcome by fixing the image of the satellite on the photograph. The camera can be moved about the axes of its mounting, so as to follow the satellite's apparent motion. On the other hand the stars then appear to move, and therefore their apparent brightness on the photograph is diminished, while that of the satellite is reinforced. The shutter then interrupts the stellar tracks at specific moments, and these breaks can be measured.

3. *Interferometric Measurements*

Optical measurements are not the only ones: radio astronomers also can track satellites. When the satellites carry transmitters they can be localized interferometrically, using their hertzian signals. While this technique is less accurate than photographic determinations, it allows us to study satellites which are too faint for optical tracking, and we can follow them under circumstances which severely hamper optical observation, as when the satellite passes through the earth's shadow or is hidden by cloud cover.

B. RADIAL VELOCITY DETERMINATIONS

Both the position and the velocity of the satellite are functions of time. Given the measurement of the velocity, theoreticians can determine the orbit. This type of measurement can be made on satellites carrying a stable narrow-band fixed-frequency transmitter, or else a reflector capable of returning to earth a measurable fraction of the radiation of a terrestrial transmitter (which could be a laser). As is well known, if v_R is the satellite's radial velocity, and f the frequency of the transmitter, the frequency shift measured on the ground is, to a first approximation:

$$\mathrm{d}f = f(v_R/c).\tag{74}$$

For a beam sent from the ground and reflected by the satellite, the frequency shift is:

$$\mathrm{d}f' = 2f'(v_R/c).\tag{75}$$

In fact it is not the instantaneous velocity that is measured, but rather the change in the distance of the satellite over the finite integration time of the measurement.

These methods are highly accurate and are already used in maritime navigation (transit system). Their accuracy is of the order of 20 m.

C. DISTANCE DETERMINATIONS

Using the newly developed *laser* technology, distances can now be measured within an accuracy of 1 or 2 m. This technique is extremely promising, and will lead to studies far more precise than with other methods.

Other systems (such as the American SECOR system) proceed to distance (and radial velocity) measurements by 'transponders' installed in the satellite.

D. ANALYSIS OF THE MEASUREMENTS. RESULTS

There is no particular problem in the reduction of the data. For optical data, it is carried out by the most traditional methods, requiring in particular correction for refraction effects. In the same way, radio measurements must be corrected for ionospheric refraction, although these determinations can be simplified by using two frequencies.

However, if one wishes to use these data for the determination of orbits and trajectories, an extended network of ground observers must be set up. Then the determination of trajectories in space reduces to purely trigonometric and geometric operations.

With the results, perturbations of the Keplerian orbits can be studied. Here we must note that the earth's rotation about its axis raises a problem: while a Keplerian orbit always lies in a plane, this plane is fixed in a coordinate system fixed in space by astrometrists with respect to the most distant objects – stars, galaxies. On the other hand, this plane is in motion relative to the geographical coordinate system. The orbit *precesses*. While this phenomenon is relatively simple for a circular orbit, it becomes more complicated for an ellipse: since near the perigee, because of the law of areas, the satellite is faster than near the apogee, the central projection of the trajectory on a map of the earth is no longer regular as for a circular orbit (Figure 15), but rather more complicated.

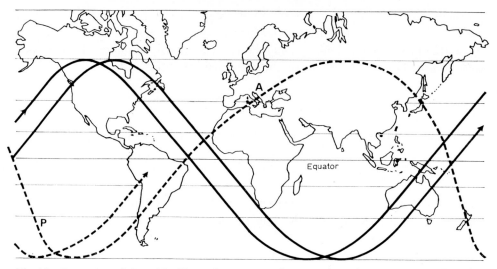

Fig. 15.   Precession of the orbit. The node moves to the west, along the equator; note how the projected orbits are distorted, as a result of their ellipticity in space and the law of equal areas. Solid line: circular orbit; broken line: elliptical orbit.

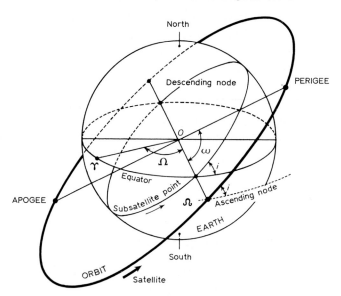

Fig. 16.   The parameters of the orbit.

Once we have resolved these difficulties, analyzed the observations, and plotted them in a fixed coordinate system, what do we find? In other words, how can we describe the orbit?

Figure 16 shows the *elements of the orbit*: the perigee and its argument; the apogee; the inclination of the orbit to the earth's equator, and finally the right ascension of the ascending node, measured from the point ♈, which is the vernal equinox.

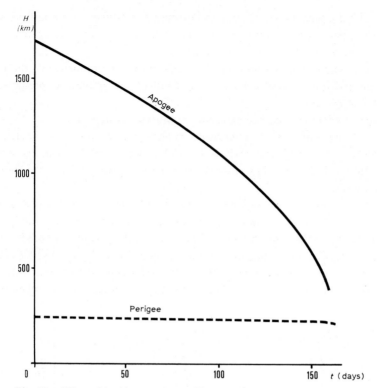

Fig. 17.   Effect of braking on the satellite Sputnik 2. (See also Figure 13.)

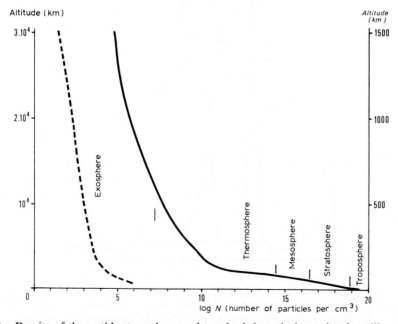

Fig. 18.   Density of the earth's atmosphere as determined through the study of satellite motions. The curve at the left refers to the scale at the left, while that on the right refers to the scale on the right. There is a single scale for the abscissae.

These orbital elements would remain constant for perfect Keplerian orbits. They change when perturbations are present. Their time evolution can then be used to study the perturbations which cause the changes.

Figure 17, which refers to Sputnik 2, clearly shows the influence of the forces of atmospheric friction. In particular the rapid fall of the apogee, as predicted above (page 34), can be seen.

This type of effect can lead to extremely important conclusions.

First of all, the *density of the atmosphere* at very great altitudes can be determined. Figure 18 shows the results which have been obtained from observations of artificial satellites. In another work, *Les observatoires spatiaux*, we shall go into more detail on this question.

It is very striking to see that the density of the atmosphere is sensitive to solar activity. Figure 19 shows a remarkable example, according to Barlier and Chassaing. Another property of the orbit, the rate of variation of the period, is plotted, for several satellites, as a function of time; we see that on May 26 and May 31, 1966, there was a strong correlation between geomagnetic index (characteristic of solar activity) and the rate of variation of the period. There are many examples of such phenomena.

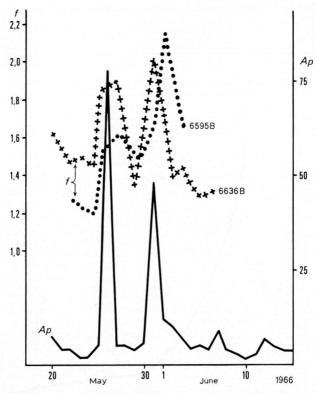

Fig. 19. Correlation between satellite motions and solar geomagnetic activity (according to Barlier and Chassaing). The geomagnetic index is A_p. For two satellites, the quantity f (which is the factor by which the model densities must be multiplied) is closely correlated with A_p.

Perturbations due to *radiation pressure* do not act in the same way as frictional forces; they have a seasonal character. From the study of such data it was possible to learn how Echo 1 lost the gas it contained (because of meteoroid impacts) and therefore how its mass varied; also the curve of the density of the atmosphere was filled in between 900 and 1500 km.

Different problems are raised by the perturbations due to the gravitational attractions of the sun, the moon, and the aspherical earth: because of the sophistication of present methods of calculation in this field, these can be computed *a priori* with high accuracy.

As a result, by the use of satellites one can go still further than the already very accurate methods of astronomy and geodesy in the determination of the relevant constants. Of course this type of research is not spectacular. It remains fundamental. The ever more precise determination of the properties of the geoid – the pear-shaped figure of our earth – amply justifies these exploits of observational and computational accuracy.

CHAPTER III

INITIATION IN ASTRONAUTICS

1. Transfer Orbits

From the summary description we have given above of Keplerian orbits we can see, first of all, that it is practically impossible to put a device in a satellite orbit directly from the ground; in Keplerian mechanics the only possibility would be a horizontal launch; obviously the topography and the atmosphere raise insurmountable difficulties, and such a launch is merely a theoretical notion.

Thus the satellite must be 'launched' from a non-negligible height H. It must therefore first be transported to this altitude and given a sufficient velocity at this altitude: this is the role of *launch rockets*. The study of the trajectories of these launch vehicles is one of the fundamental problems involved in placing any type of satellite in a Keplerian orbit. This is the problem of *transfer orbits*.

The problem is a celestial mechanics problem whose constraints are defined by requirements of economy: either the energy needed for the transfer or the time involved must be minimized. Moreover the effects of possible guiding or pointing errors must be considered: the greater these errors can be, the higher will be the load, since fuel must be carried for in-flight corrections of the trajectory. This necessary faculty of corrections must be underlined: with controlled flight we have left celestial mechanics for astronautics. It is not enough to leave things up to the gravitational field (one of my colleagues put it: "to fall asleep in Newton's arms"). Just as for any terrestrial vehicle, *navigation* is necessary and from then on celestial mechanics is no longer of much use, except for the reduction of fuel consumption by 'turning off the engine', somewhat like the stingy driver who turns off the ignition of his car after crossing a pass, descending according to the law of gravity alone (although against most highway laws).

The energy consumption at lift-off is proportional to v_0 and naturally to the mass (satellite, plus launch vehicle, plus fuel) but does not depend on the launch angle V_0. Obviously, if we simply want to go from one point to another, the ellipse is the most economical trajectory (Figure 20). A parabolic (or higher) velocity would give enough energy to the missile for it to reach 'infinity'. If such an initial velocity is used to reach a finite distance, the missile will retain a useless residual energy.

Kepler's Third Law gives the travel time from pericenter to apocenter (half of the period):

$$t_1 = \frac{T}{2} = \frac{\pi}{2\sqrt{2}} \frac{(r_{min} + r_{max})^{3/2}}{GM_\oplus}. \tag{1}$$

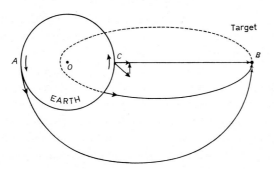

Fig. 20.   There are several paths possible between the earth and the target.

However, the transfer vehicle might be launched from any other point of the earth's surface. A vertical shot would give an extreme value which (assuming zero velocity at the target B) is:

$$t_2 = \frac{1}{(2GM_\oplus)^{1/2}} \int_{R_\oplus}^{r_{max}} \left(\frac{1}{r} - \frac{1}{r_{max}}\right)^{-1/2} dr \tag{2}$$

or

$$t_2 = \frac{(r_{max})^{3/2}}{2(2GM_\oplus)^{1/2}} [\pi - \theta_0 + \sin \theta_0], \tag{3}$$

where:

$$R_\oplus = r_{max} \sin^2(\theta_0/2) \tag{4}$$

defines θ_0.

A simple calculation shows that for a shot to the moon:

$$t_1 \sim 120 \text{ hours}$$
$$t_2 \sim 116 \text{ hours}.$$

In each of these cases the initial velocity can be calculated. The values found are very slightly lower than the parabolic velocity, since the moon is very distant.

Vertical launch: $v_0 = 11.1 \text{ km sec}^{-1}$
Horizontal launch: $v_0 = 11.1 + 1.5 \text{ km sec}^{-1}$.

However, it must be remembered that the rotation of the earth imparts a motion to the launch point whose velocity (0.46 km sec^{-1} at the equator) must be added (vectorially) to the launch velocity. In fact, for a vertical launch, the initial velocity must be suitably tilted and slightly increased (Figure 20, point C).

For a horizontal shot, the launch velocity can be reduced by 0.46 km sec^{-1}. However, the circumstances are this favorable only if the moon can be reached at the moment it crosses the plane of the earth's equator.

We see that theoretically a transfer orbit with horizontal launching is somewhat advantageous. In practice this is not so: the launch from the earth is always perturbed by the frictional forces of the dense layers of the atmosphere. These forces (which obviously lead to an increase in the launch energy necessary, since they act as braking forces) are far greater for horizontal firings.

The travel time is naturally not the only quantity involved: this would be true for a trip from one fixed point to another; in most cases, however, the problem is not only to place the satellite at the proper initial altitude, but also to give it the correct initial velocity, so that it will then be easy to bring its trajectory onto the desired orbit with a low fuel consumption. If this orbit is to be followed in the same direction as the earth's rotation, it comports two advantages, since the satellite benefits from the earth's rotation both at launch from the surface and at injection in the orbit, inasmuch as the tangential component is maximum. The areal law gives the velocity:

$$v_{r_{max}} = \frac{r_{min}}{r_{max}} v_0 . \tag{5}$$

At the distance of the moon (whose orbital velocity about the earth is 2.39 km sec^{-1}) the velocity $v_{r_{max}}$ will be 0.19 km sec^{-1} and the relative velocity at impact will therefore be quite large (necessitating the use of retro-rockets to slow down and brake the vessel landing on the moon). This relative velocity would be nearly the same for a vertical launch or for a horizontal launch in the opposite direction. But while this is true for the moon it is not so for a lower orbit.

In order to reach a circular orbit at altitude H, using a perigee-apogee type transfer, we have:

$$v_{r_{max}} = \left(\frac{R_\oplus}{R_\oplus + H} \right) v_0 , \tag{6}$$

while the velocity which must be reached should be at least the circular velocity (it would hardly be of interest to send the satellite back downwards, which is why we say 'at least'), i.e.:

$$v_{r, \text{circ}} = \frac{2\pi (R_\oplus + H)}{P} = \left[\frac{GM_\oplus}{R_\oplus + H} \right]^{1/2} \sim \frac{1}{R_\oplus} \left(1 - \frac{H}{2R_\oplus} \right) (GM_\oplus)^{1/2} . \tag{7}$$

The height H of the apogee determines v_0:

$$v_0^2 = \frac{2GM_\oplus}{R_\oplus} - \frac{GM_\oplus}{R_\oplus + H} , \tag{8}$$

i.e.

$$v_0^2 \sim \frac{GM_\oplus}{R_\oplus} \left(1 + \frac{H}{R_\oplus} \right) \tag{9}$$

and

$$v_0 = v_{\text{circ}, 0} \left[1 + \frac{H}{R_\oplus} \right]^{1/2} . \tag{10}$$

This velocity is called the *Hohmann transfer velocity*.

Consequently the velocity needed for transfer (*characteristic velocity*) must be at least:

$$v_{\text{char}} = v_0 \cong v_{\text{circ},0} \left[1 + \frac{H}{2R_\oplus} \right]. \tag{11}$$

This is the satellization velocity shown on Figures 5 and 24. Transfer trajectories of this sort were studied by Hohmann, and have been given the name 'Hohmann semi-ellipses'. In fact, when we consider second-order terms, the velocity $v_{r,\max}$ they give to the apogee is no longer equal to the circular velocity at height H. More rigorously we find:

$$\frac{v_{r,\max}}{v_{r,\text{circ}}} = \frac{R_\oplus}{R_\oplus + H} \frac{v_{\text{char}}}{v_{\text{circ},0}} \left(\frac{R_\oplus + H}{R_\oplus} \right)^{1/2} \sim 1 - \frac{H}{R_\oplus}. \tag{12}$$

Therefore at the moment of 'injection into the orbit', a supplementary velocity $v_0 H/R_\oplus$ must be imparted to the satellite. If $H/R_\oplus = 0.1$, this velocity remains low, of the order of 1 km sec^{-1}.

Thus the Hohmann semi-ellipses are quite notably economical – but once again, air resistance renders them totally useless.

We might consider what appears to be an even more economical orbit: close to the Hohmann orbit, but not tangent to the desired orbit, this orbit would have an apogee slightly more distant from the earth than the orbit to be reached. A slight increase in v_0, by a quantity of order $v_0 H/R_\oplus$, would do. However, this can be excluded, for at the moment of injection into orbit, the direction of the satellite's motion would have to be changed by an additional centripetal radial velocity so as to bring it onto the circular orbit. Calculation shows that one loses more than one gains!

Might there be other transfer orbits, tangent to the orbit to be reached, but corresponding to non-horizontal launch velocities? It is easy to show that the velocity would then be somewhat smaller than for the Hohmann ellipse, and consequently the apogee velocity correction would have to be somewhat larger in order to put the satellite into a circular orbit. This is quite clear when one considers that such a transfer orbit has the same apogee as the Hohmann orbit, but a perigee closer to the earth's center.

Once again, Hohmann orbits run into the serious problem of air resistance. Thus a different type of orbit is to be preferred.

First of all, there is the *vertical firing*, which minimizes air resistance problems, but complicates those of the injection into orbit. The so-called 'synergic' orbit is obtained by starting with a vertical firing and then applying in-flight corrections to the trajectory so as to make it horizontal at the desired altitude. Such orbits (proposed by Oberth) are computed by setting the rate of energy loss by air resistance equal to the rate of increase of potential energy. Obviously they depend on the shape of the launch vehicle.

Other types of orbits, based on *in-flight modifications of the vehicle's energy*, can be devised. They have their advantages, but naturally require more fuel. It has been shown that even if one considers orbits of continuously increasing energy, the Hohmann orbit is still theoretically the most economical (neglecting air resistance). However, another serious defect of Hohmann trajectories must be underlined. They necessitate a sudden change in velocity, and therefore strong accelerations, which are dangerous for both instrumentation and any astronauts on board. Moreover it is important to remember that a small firing error leads to large errors at the apogee, so that the faculty of in-flight corrections remains necessary in any case.

This sort of argument shows why multi-stage rockets are used: with these, dead weight can be left behind while the energy is being increased, so that the energy gain works more efficiently in increasing the orbital velocity.

To conclude, I shall discuss a question of terminology. Many purist scientists are dismayed to read or hear in the mass media that 'such a satellite has *reached* its orbit, or trajectory'. Of course this may seem silly, since any missile is always, from firing to landing, or destruction, on its trajectory or on its orbit. Of course what is meant (though not always stated this way) is that a certain missile, in the course of its trajectory, has moved from one segment (where e.g. its energy was being modified by the action of the rocket engines) to another. In general, one usually means: 'to another segment where gravity alone acts' – or in other words 'to a Newtonian segment of its orbit, or trajectory'. The word 'orbit' is the proper one for a closed path which the device follows several times in succession; 'trajectory' is a more general one.

2. Toward Venus and Mars

What we have said concerning the voyage from the earth to the moon (or to a lower circular orbit for an earth satellite) may be applied *mutatis mutandis* to the launching of a vessel from the orbit of the earth around the sun to the orbit of another of the sun's planets such as Mars or Venus. From the point of view of celestial mechanics, this problem is more complicated than that of an earth satellite. In determining the trajectory, we must first solve the Keplerian problem of earth plus rocket; a sun-plus-rocket Keplerian problem joins onto this, and the transition goes by way of a three-body theory; upon arrival near Mars, an analogous transition transforms the problem into the Keplerian rocket-plus-Mars configuration (and the same is true for Venus or any other planet), and all this is independent of non-Newtonian in-flight corrections.

In view of the complexity of this problem, we certainly cannot give the fully general solution here. However, a number of nearly intuitive points can be easily understood, giving at least an idea of the magnitudes involved, as a basis for further discussion.

First of all the launch velocity must be close to the parabolic velocity if not higher (hyperbolic): in that case with a slight additional energy it is possible to reach the Martian orbit, which is nearly at infinity as far as the earth's gravitational field is concerned.

In the rocket-plus-sun configuration, the initial velocity of the probe will in fact be the velocity at infinity of the hyperbolic trajectory relative to the earth, to which the earth's orbital velocity has been added. For a parabolic trajectory the velocity at infinity would be zero; thus an object launched at parabolic velocity from earth will stay close to the earth's orbit. The only difference between the two orbits would arise from the difference of the barycenters of the probe-sun and earth-sun systems (i.e. in fact from the influence of the earth on its own orbit).

However, the transfer orbit must be chosen correctly. Once the problem of leaving the earth has been dealt with, the initial velocity must place the probe in a Hohmann orbit. If we assume that the probe has already been brought to the altitude H, it can orbit there before being launched toward Mars, it can be launched immediately, or else it can be launched from an intermediate station in another circular orbit: in any case we can assume that the problems of air resistance have been overcome. Once the height H has been reached, the Hohmann orbit is the most appropriate one, and the launch velocity needed is easy to compute: it is the sum of the velocity for a nearly parabolic orbit in the earth-linked system and a characteristic velocity which in the solar system is quite different from the parabolic velocity. In the solar system, at the earth's orbit, the parabolic velocity (see above, p. 20) is:

$$v_p = \sqrt{2GM_\odot/D_\oplus} \sim 42.1 \text{ km sec}^{-1}, \tag{13}$$

while the characteristic velocity of Hohmann earth–Mars trajectory is equal to:

$$v_{\text{char}} = \sqrt{\frac{2GM_\odot}{D_\oplus} - \frac{GM_\odot}{D_{\text{Mars}}}} \sim 3.65 \text{ km sec}^{-1}. \tag{14}$$

At the orbit of Mars the velocity relative to the sun will be:

$$v = (v_{\text{char}} + v_\oplus) \, D_\oplus/D_{\text{Mars}} \sim 21.5 \text{ km sec}^{-1}, \tag{15}$$

while the orbital velocity of Mars is:

$$v = 2\pi \frac{D_{\text{Mars}}}{T_{\text{Mars}}} \sim 24 \text{ km sec}^{-1}. \tag{16}$$

Thus the relative velocity will be 24.0 ± 21.5 km sec^{-1}. The minus sign corresponds to the more favorable case: then $v \sim 2.5$ km sec^{-1}.

Note that the probe still has to enter the gravitational field of Mars and land on the planet. If we assume that the probe passes fairly far from Mars, at a distance where the parabolic velocity is less than 2.5 km sec^{-1} (i.e. an altitude greater than three Mars radii), its motion must be modified. At the surface of Mars the parabolic velocity is 5.15 km sec^{-1} (Figure 21). The initial energy must be used up by braking; if we want to land on the surface, the braking energy must correspond to a velocity $\sqrt{5.15^2 + 2.5^2} = 5.72$ km sec^{-1}, which is just the velocity needed to launch a vehicle from the surface along a hyperbolic trajectory with a velocity at infinity of 2.5 km sec^{-1}. If we simply wish to place the probe in a very low circular orbit around Mars, it must be slowed down to a speed of 3.64 km sec^{-1} in the Mars-linked system; thus an additional velocity of $5.72 - 3.64 = 2.08$ km sec^{-1} is needed to reach this orbit.

The problem of trips to other planets is obviously the same. For inferior planets

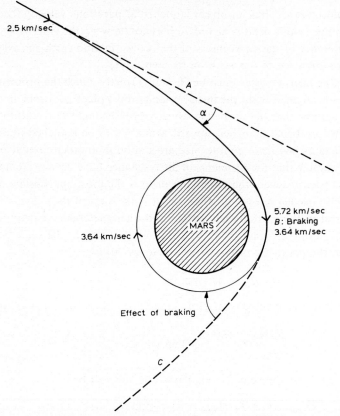

Fig. 21. Braking operations necessary to place a device in circular orbit around Mars. Trajectory *A* is nearly a straight line relative to Mars, although it is part of an ellipse around the sun. Angle α is the deviation from *A* due to the attraction of Mars. *C* is part of the hyperbolic trajectory relative to Mars that would be followed in the absence of braking.

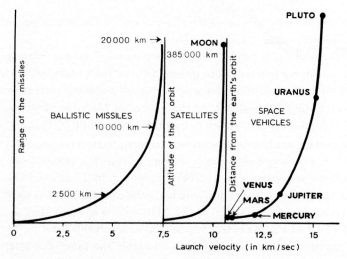

Fig. 22. Relation between the initial velocity and the range of the device.

such as Venus it can be noted that the earth–Venus voyage resembles the return trip from Mars back to earth, which can be made by reversing the velocities in the preceding example.

Figure 22 shows how the launch velocity is related to the destination chosen.

3. Employment of Double Maneuvers

There is one very general rule in this sort of problem: one cannot avoid imposing artificial accelerations, in addition to those of gravitation, at various points. This additional action should take place as close to the centers of attraction as possible: it is the kinetic energy, proportional to v^2, that is increased; for a given velocity increment the increase in kinetic energy will be greater the closer one is to pericenter, where the law of areas shows that the velocity is maximum.

Let us go into this point with an evaluation of the 'efficiency' of the operation. Let v_1 and v_2 be the velocities before and after the impulse. The energy difference is:

$$\delta E = \tfrac{1}{2}m\,(v_2^2 - v_1^2) = mv\,\delta v, \tag{17}$$

since the potential energy has not changed.

However, the amount of energy to be furnished is only:

$$\delta' E = \tfrac{1}{2}(v_2 - v_1)^2\, m = \tfrac{1}{2}(\delta v)^2\, m. \tag{18}$$

The efficiency is therefore:

$$\frac{\delta E}{\delta' E} = \frac{v_1 + v_2}{v_2 - v_1} \sim 2\,\frac{v}{\delta v} \tag{19}$$

given δv or $\delta' E$, the efficiency and δE increase as v increases. Because of this, methods taking advantage of this phenomenon and using *double thrusts* have been suggested; we shall give an illustration of the concept using an example due to Berman.

Let us consider the following problem: A device is in a circular orbit, and we wish to send it to infinity along a parabolic or hyperbolic trajectory. The simplest solution *a priori* is given by a trajectory of type (a) (Figure 23), where artificial acceleration

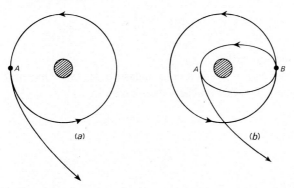

Fig. 23.   How to leave a circular orbit in order to reach infinity.

occurs at point A alone. Another solution, of type (b), exists, where the device must be propelled at two points A and B.

In maneuver (a), we have:

$$\delta v_A = v_{\text{circ}} - v_{\text{hyp}}. \tag{20}$$

In maneuver (b), we have:

$$\delta v_B = (v_{\text{circ}} - v_{\text{ell}, B}) + (v_{\text{hyp}} - v_{\text{ell}, A}). \tag{21}$$

The energy of the missile in its hyperbolic orbit is equal to:

$$E = \tfrac{1}{2}mv_{\text{hyp}}^2 - E_{\text{pot}} = \tfrac{1}{2}mv_\infty^2 \quad \left(E_{\text{pot}} = \frac{GM_\oplus}{r_A}\right), \tag{22}$$

and so we deduce:

$$v_{\text{hyp}} = \sqrt{v_\infty^2 + \frac{2GM}{mr_A}}. \tag{23}$$

We know what the various velocities in play are: the velocities in elliptical orbit at A and B are

$$v_{\text{ell}, A} = \sqrt{GM\left(\frac{2}{r_A} - \frac{2}{r_B + r_A}\right)} \tag{24}$$

and

$$v_{\text{ell}, B} = \sqrt{GM\left(\frac{2}{r_B} - \frac{2}{r_B + r_A}\right)}. \tag{25}$$

The circular velocity is: $v_{\text{circ}} = \sqrt{GM/r_B} = v_p\sqrt{2}$.

Thus the velocity increment involved in a maneuver (b) is (as can easily be verified);

$$\delta v_{\text{b}} = v_{\text{circ}}\left[1 - \sqrt{2\left(\frac{r_B}{r_A} + 1\right)} + \sqrt{2\left(\frac{r_B}{r_A} + \frac{v_\infty^2}{v_p^2}\right)}\right]. \tag{26}$$

On the other hand, for the simpler maneuver of type (a), we have:

$$\delta v_{\text{a}} = v_{\text{circ}}\left[-1 + \sqrt{2\left(1 + \frac{v_\infty^2}{v_p^2}\right)}\right]. \tag{27}$$

using the same relation with $r_A = r_B$.

The quantity $\Delta = \delta v_{\text{b}}/\delta v_{\text{a}}$, which is a measure of the 'efficiency' of the double maneuver, obviously depends on the values taken by:

$$\alpha = 2\left(1 + \frac{v_\infty^2}{v_p^2}\right) \tag{28}$$

and

$$\beta = 2\left(1 + \frac{r_B}{r_A}\right). \tag{29}$$

We have:

$$\Delta = \frac{1 - \sqrt{\beta} + \sqrt{\beta + \alpha - 4}}{\sqrt{\alpha - 1}}. \tag{30}$$

Several cases may be considered: first of all if $v_\infty/v_p \sim 0$ (the final trajectory is parabolic), $\alpha = 2$. The efficiency of the double maneuver depends on β alone:

$$\Delta = \frac{1 - \sqrt{\beta} + \sqrt{\beta - 2}}{\sqrt{2} - 1}, \tag{31}$$

which increases as β increases. As r_B/r_A goes from one to infinity, Δ goes from 1 to $1/(\sqrt{2} - 1)$.

As α increases (hyperbolic final trajectory) the efficiency of the double maneuver, at equal values of β, decreases. As α tends to infinity, Δ tends to 1, for all values of β.

Thus a double maneuver of type (b) is always advantageous, the more so when the final velocity approaches the parabolic velocity, and the intermediate orbit has the lowest possible perigee A; the only limits come from other factors, mainly atmospheric friction, which excludes launch points too low in the atmosphere.

4. Complex Orbits

Obviously problems of this type can become extremely complex, treated as an n-body problem, and taking account of the rotation of the planets and the 'planetocentric' latitude of the missiles at each moment. Just as an example, we shall mention the geocentric trajectory of Pioneer 4, which is familiar to astronomers since it appears regularly on the covers of the proceedings of the annual COSPAR meetings; also

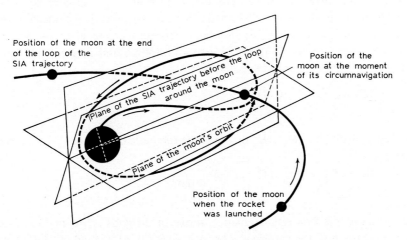

Fig. 24.   Trajectory of Lunik 3 in a coordinate system referred to the position of the moon at the moment of its circumnavigation. (Courtesy of the Soviet Information Office.)

Figure 24 shows the trajectory of Lunik 3 which flew around the moon. Naturally such trajectories take a simpler form in heliocentric coordinates: this, e.g., is the case of Lunik 1 (Figure 25).

It is impossible to go into more detail here.

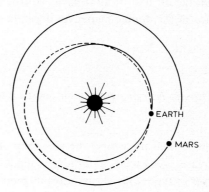

Fig. 25.   The projection of the trajectory of Lunik 1 on the plane of the ecliptic (according to Danjon and Muller).

5. Effects of Firing Errors

We have already mentioned the important problem of firing errors which necessitate in-flight corrections; these may have to be continuous, and must be commanded from the ground or, in the case of a manned vehicle, by the astronaut. Naturally the calculation of possibly necessary corrections must be prepared in advance of each launch. Some trajectories will require only slight modifications; even for comparable initial errors, others may require considerable corrections. We cannot discuss the problem in all its generality here, but we can give a simple example. Let us consider a satellite launched from earth which we wish to place in a circular orbit. If the injection velocity and the injection angle (from the desired horizontal) are in error by 1% and 1° respectively, what will be the resulting errors in apogee and perigee? The injection velocity can be obtained from the equation:

$$\frac{2GM_\oplus}{R_\oplus + H} - v_0^2 = \frac{GM_\oplus}{R_a}. \tag{32}$$

For a circular orbit, $R_a = R_\oplus + H$; this fixes the theoretical value of $v_0 = v_{\text{circ}}$. We then see that:

$$\frac{R_a - (R_\oplus + H)}{R_\oplus + H} = \frac{\delta v_0^2}{v_0^2} = \frac{2 \, \delta v_0}{v_0} = 0.02. \tag{33}$$

Inasmuch as the injection point remains on one of the ends of the major axis, if the velocity is 1% too low, the perigee will be lowered by 2% (for small H, about 140 km). This shows quite well that very high accuracy is needed. The error in angle results in an error in the eccentricity e. In the case above V_0 in principle was computed from:

$$e^2 = 1 - \sin^2 V_0 = \cos^2 V_0 . \tag{34}$$

Consequently

$$de = - \sin V_0 \, dV_0 \tag{35}$$

and

$$\left| \frac{de}{e} \right| = \frac{\sin V_0}{\cos^2 V_0} \, dV_0 . \tag{36}$$

In the neighborhood of $V_0 = \pi/2$, we have

$$\sin V_0 \sim 1, \quad \cos^2 V_0 \sim \left| \frac{\pi}{2} - V_0 \right|^2$$

and the error is very low.

In the ideal case, $e = 0$; an error of $1°$ leads to $e = 0.0003$. Then the error in r_{max} and r_{min} is:

$$(R_\oplus + H) \, e \sim 0.00029 \, R_\oplus .$$

Both apogee and perigee are changed by 0.03%; the error is about 2 km.

Clearly such calculations are vital when launches toward the moon, Venus, or Mars are in question: the errors can become very significant, the more so when the target is more distant.

Still, we have already attained a remarkable mastery of this technique: Table II, pp. 9–11, with its impressive list of successful experiments, shows this well.

6. Conclusion

The biggest computers of the space centers of the U.S.S.R. and the U.S.A. (and also Europe – and soon other countries) work incessantly on the calculation of orbits and the preparation of future launches. Elaborate calculations continue, piling up tons of punched cards and magnetic tapes. Note that such calculations treat all the perturbations very carefully, naturally taking into account the possible modification of the plane of the orbit with each perturbation, a further complication of the problem. The reader will understand that we have only attempted to orient him in this branch of the science. Thanks to artificial satellites and planets, the celestial mechanics of Lagrange and Le Verrier have a splendid field of application, requiring both subtlety and precision, and in which calculation paves the way for the distant expeditions of the Magellans of space. It is a striking example of the fact that the 'utility' and 'efficiency' (so dear to budget administrators) of fundamental research can only be evaluated over very long periods of time. Would Lagrange have received the laboratories, funds and respect that the NASA in 1967 accords its mathematicians? Without any doubt, no; besides, that was not the problem. Still, the lesson should be taken serious by those of today's governments hesitant in their support of pure research, and overeager to plan 'useful' research.

CHAPTER IV

WHAT OF EXPERIMENTAL ASTROPHYSICS?

The study of the motions of satellites and space probes indeed follows the methods of celestial mechanics, and gives us information on the physics of the forces acting on these devices – and as we have shown, not only gravitational forces. However, these artificial celestial bodies do not actually resemble real celestial bodies as far as their own physics is concerned. Now while it is important to know how the external medium acts on heavenly bodies, it remains equally necessary to learn how these bodies react to these outside influences, which condition their very nature. This is a first reason for attempting to 'produce', within their natural medium, artificial celestial bodies similar in nature to real objects. Of course these cannot be stars or planets. On the other hand it seems natural to produce artificial *comets* and artificial *meteoroids*. Indeed, real comets and meteoroids occur in the immediate vicinity of the earth, and even enter the atmosphere, where they can be destroyed by explosion or volatilization.

By producing false comets and false meteoroids in this way, we can hit two birds with one stone. We can study the reaction of the object to the environmental conditions; repeating the experiment several times, we will become familiar with its own reactions, more familiar than would be possible with natural objects, whose nature is not well known and moreover whose appearance is generally unexpected. So, to begin with, we advance our knowledge of the environment.

Besides this (and perhaps more importantly!), the nature of true comets, or of true meteoroids, becomes clearer, and our knowledge advances where before it was imperfect. This is a major step.

The first realistic proposal of experiments of this type was no doubt that of F. Zwicky. It is well known how many extremely interesting results have been obtained from the study of real meteoroids by space devices. Zwicky's proposal was to produce artificial meteor showers. The first attempts in 1946 were unsuccessful, but later experiments succeeded in 1957 – and since then other authors have continued analogous work. The problem is to eject self-luminous objects from the head of the rocket. The reaction of aluminum with ferric oxide ($2 Al + Fe_2O_3$) was suggested for this purpose. Ejection from the rocket was at an altitude of 60 km. With this sort of experiment one can in principle verify the theory of atmospheric braking of natural meteors and meteorites, as well as the theory of their volatilization. The minimum dimensions of observable meteors can also be estimated from the experiment, and this clarifies the interpretation of observations of real meteors.

In fact what is involved here is the production and study of an artificial 'meteor'

Fig. 26.   Launching artificial meteors. (Courtesy of Zwicky.)

phenomenon, rather than artificial 'meteoroids'; the nature of the object itself is hardly involved (see Figure 26).

We are far less advanced where comets are concerned. We have to recognize that comets are rare objects, while meteoroids fill interplanetary space. Space probes can study the distribution and properties of meteoroids quite well; however, while the physical exploration of a comet by a nearby passage of a space probe has been envisaged, it has not yet been attempted, and it will no doubt be difficult.

The fabrication of an artificial comet can also be envisaged. Among the problems of comet physics which might be attacked by such a project, the following can be mentioned:

When comets (whose orbit is highly eccentric) approach the sun, the gases they contain are ejected, and the rapidly changing physical and chemical conditions to which the comet is subjected govern the formation of various ions and free radicals in the ejected gas. The brightness of the head of the comet also depends on these conditions. An artificial comet, 3 m in diameter, located in a stationary circular orbit ($H \sim 36\,600$ km)* would have a head $\frac{1}{2}°$ in apparent radius, and no doubt be of 10th magnitude. It seems then that the experiment is quite feasible. Some sort of block of ice put in such an orbit would have a lifetime of several days according to some authors; others argue that it would be destroyed too rapidly for the experiment to be worthwhile.

Preliminary to such research, it has been possible to create artificial clouds of ions in the interplanetary medium. The ejection of alkali metals, a type of experiment quite current in geophysical applications, has been considered. Ammonia (NH_3) clouds have already been ejected at 200 km altitude. From the photographs and spectra obtained (see e.g. Wurm, Rosen) it has been shown that within the expanding layer of the cloud, no dissociation of the ammonia molecule NH_3 due to solar ultraviolet radiation could be observed, contrary to what seems to occur in comet tails. From this negative result one could assign definite limits to the dissociation probability of NH_3 by solar ultraviolet radiation, as well as to the excitation probability for the fluorescence of NH_2 in solar radiation, or rather to the product of these two quantities: there may have been dissociation, but without the appearance of the resonance spectrum of the radical NH_2 between 5000 and 7000 Å. The ejection of other gaseous molecules such as CO_2 and water vapor H_2O has also been envisaged.

It is probable that there will be considerable development of this type of research in the future. Comets in fact are doubly interesting in that they are both interesting physical chemical objects, and they yield information on solar activity, e.g. by way of the 'solar wind' of corpuscles which influences the shape and development of comet tails. In spite of the difficulties involved in the experiments, they are well worthwhile.

---

* This orbit has no resemblance whatsoever to real comet orbits; it simply allows us to study the behavior of the comet, as though it were a real comet at the same distance from the sun as the earth.

CHAPTER V

THE DIRECT EXPLORATION OF THE
EXTRATERRESTRIAL WORLD

1. The Direct Exploration of the Moon

Table II (pp. 9–11) summarizes the major phases of the exploration of the moon: lunar impact (Luna 2), photographs of the invisible side (Luna, Lunar Orbiter, etc.), high resolution photographs (the Ranger series), close-up photographs of the surface itself (Zond 3), circumlunar orbits (Luna 10, Lunar Orbiter, etc.), attempts at analyzing the mechanical and chemical properties of the lunar soil (Surveyors), and most recently the manned landing on the moon (Apollo XI) with the return of samples of lunar material, as well as many other experiments. Obviously between the time this paragraph is written and the time it is read, other experiments will have been carried out, new data acquired, so that much of what is written below may be out-of-date.

What are the problems that have been solved by the success of these launches? Also, what are the problems that can be studied by future launches, or at least by the nearest ones to come?

We know that the moon is one of the largest satellites in the solar system. While it is no longer our only satellite (since there are so many artificial ones!) it is at least by far the biggest and most notable one. Nevertheless its mass is too small to have retained an atmosphere, if ever one existed. This conditions the appearance of the moon, traditionally considered to be a 'dead' body. Galileo discovered its relief; starting with his observations, a regular lunar cartography has developed, and continues to be elaborated (from earth observatories) to satisfy the rebirth of interest in lunar problems stimulated by progress in astronautics. The optimum resolving power (limited by the earth's atmosphere and by the size of the instruments) used to be 1 or 2 km on the moon; smaller details were unobservable. By measurements of the radiation reflected by the moon (eclipses, infrared, radio region), the temperatures and the conductivity, as well as (by polarization) the microscopic structure of the surface, can be determined. Photometry of craters is involved in the detailed topographic study of the moon. Where astrometry and celestial mechanics are concerned, the moon, as the only natural earth satellite, is a body of considerable importance: the study of its distance and its orbit yields a determination of the sun-earth mass ratio; the dynamical theory of the motion of the moon involves the problem of perturbations by a third body (the sun); finally the theory of tides is closely related to the study of the moon's orbit, and the precession of the equinoxes is also a result of the moon's attraction on the earth.

The origin of the moon and its relief has always been a problem of interest to theoreticians: Was the moon ejected by the earth? This hypothesis appears impossible according to the laws of mechanics. It seems more likely that the moon was formed by a cumulative process of agglomeration within the dust cloud surrounding the sun during its birth. Are the craters due to volcanic eruptions? Spectroscopic observations by Kozyrev have shown a weak but measurable emission in the Alphonsus crater, extending over wavelengths longwards of a sharp cut-off at 4740 Å (Figure 27). This emission is attributed to a band of the carbon molecule C_2. Several eruptions of this type have occurred in the Alphonsus crater, and this is a further argument in favor of

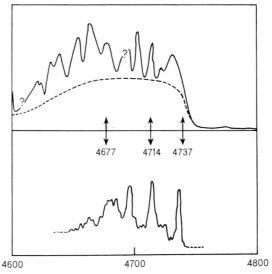

Fig. 27.   Emission spectrum observed in the Alphonsus crater (according to the work of Kozyrev). Above: Kozyrev's observations; below: molecular band spectrum of carbon C_2. The rather good agreement of the emission peaks is notable.

the *internal* origin of these clouds of soot. Is this true volcanism? That is not certain: in fact the soot is not hot, and the analysis shows that it shines by fluorescence alone (and these measurements remain highly controversial!). Many astronomers argue strongly in favor of a 'meteoritic' origin; for the most part they refer to the evidence of laboratory experiments, in which one attempts to reproduce crater formation by the impact of various materials fired onto different targets... The results are always convincing, indeed too convincing, for in the present state of our knowledge, they show how much the problem remains unsolved. Moreover it can be shown that the relation between the depth and diameter of a crater (Figure 28) is the same for lunar craters as for bomb craters (objects falling to the ground) on earth. On the other hand, terrestrial meteorite craters do not obey this relation. Is this an argument against the meteoritic theory of lunar craters? Probably not, since over the course of time erosion has levelled and filled up the terrestrial formations, and this is probably not the case on the moon, where erosion is far slower (Figure 28). On the other side also, the comparison of the forms and structures of terrestrial volcanism with those of the lunar relief is not very convincing. Until there is direct exploration, the question of the origin of the craters remains open: internal and volcanic? or external, meteoritic? Or are both phenomena involved?

Now let us see what answers space research has given to such questions.

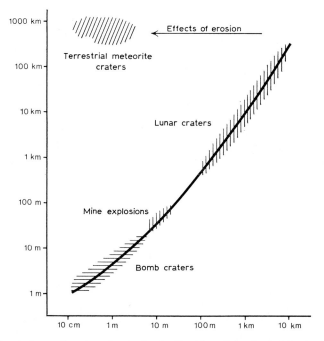

Fig. 28.   Relation between the size of a crater (ordinate) and its depth (abscissa) (according to Baldwin). The effects of erosion on terrestrial meteorite craters should be noted.

A. LUNAR TOPOGRAPHY. LARGE SCALE STUDIES

One of the first results of the Soviet and U.S. space experiments was to provide, on a large scale, a complete topography of the moon. We shall note here that contrary to the usual notion, we do not see just 50% of the moon from the earth. In fact, although it is true that the moon always turns the same side toward us, the moon appears to oscillate about its mean position. This oscillation is a combination of (a) a true oscillation, (b) (especially) the combined effect of the uniform rotation of the moon and its motion in a significantly eccentric orbit, and (c) the fact that the equatorial plane of the moon does not coincide with its orbital plane. Because of these effects (combined under the name of 'libration'), 59% of the moon is visible from the earth; however, since the regions close to the limb are observed nearly at grazing incidence, they are in fact poorly known.

The flight of Luna 3 (October 4, 1959) revealed a major portion (but not all) of the hidden 41%, including some notable features. The flight of Zond 3, and especially the succeeding flights of the first four Lunar Orbiters, brought our knowledge of the hidden side to the level of that of the visible face; Figure 29 shows an excellent photograph of part of the 'hidden' side of the moon (a small part is still unobserved). Note the striking black feature, the crater Tsiolkovsky. Other features, such as the crater Jules Verne, the Mare Orientalis with its concentric mountain ranges separated

Fig. 29.   The 'hidden' face of the moon (photograph taken by Lunar Orbiter 3, on February 19, 1967). The face of the moon which cannot be seen directly from earth was first revealed by Soviet probes in 1959. This picture shows the very dark Tsiolkovsky crater in particular.

by concentric rings of valleys, can be seen on other pictures*; craters and walled plains are very abundant.

From the first pictures published, it became clear that the topography of the two sides of the moon, that facing the earth and the other side, was quite different in character. On the other side, the pictures sent back by Zond 3 and the Lunar Orbiters reveal fewer dark *maria*, lighter colors, and a mountainous structure rich in craters.

Can these differences be explained? There has been intense discussion of what may take place on one side of the moon and not on the other: while eclipses occur only on the earth's side, and produce sudden temperature changes, they will have little influence, since the temperature changes cannot penetrate very deeply into the poorly conducting lunar soil. Another asymmetric effect, tides in the lunar crust caused by the earth's attraction, also does not seem to be very important. However, it may have been sufficient at some time in the moon's past history when the moon might have been closer to the earth than now; still, this is not certain. The shielding effect of the earth, which prevents some meteorites from reaching the visible face, is also very small: the fraction eliminated is certainly less than a thousandth. In fact, the earth is also strongly asymmetric, as is the sun and each of the planets: the observed lunar asymmetry is no doubt not excessive compared to what might be assumed *a priori* the most 'probable' asymmetry.

We shall add that a commission of the International Astronomical Union is presently at work on the official nomenclature of the numerous features recently discovered on the lunar surface.

B. FINE STRUCTURE OF THE LUNAR SURFACE

The information provided by Ranger 7, Ranger 8, and Ranger 9 is of a different nature. These rockets took a series of photographs of the moon during their fall to its surface. The observations have been widely discussed, and the major results can be summarized as follows.

The smallest details visible on these pictures correspond to a few centimeters resolution.

In addition to the craters observed from earth, there are many small craters; more and more craters are observed as the resolution is improved. Figure 30 shows how the number of craters observed is related to their size. These numbers were estimated from a sampling by Ranger 7 in Mare Cognitum, and reduced to a surface of 10^6 km^2 and a time lapse of 10^9 years (dividing the real number for 10^6 km^2 by 4.5 so as to take account of the age of the moon, assumed equal to that of the earth, namely 4.5×10^9 years). Also, those regions where rays radiate from craters like Tycho are more than ten times richer in craters than the regions between rays: this is shown in Figure 30.

Interpreting these statistics, and examining detailed cases, we see that we must distinguish between *primary* craters, which are essentially what we observe between

* See in particular the atlas of lunar photographs published by NASA.

rays, and *secondary* craters, which are found along the rays and which were probably caused by the fall of fragments previously ejected from the moon's surface at the primary craters. In general, primary craters are sharp, steep, and nearly circular; secondary craters are clearly elongated, less sharp, more irregular and often complex in structure, and are associated with other secondaries.

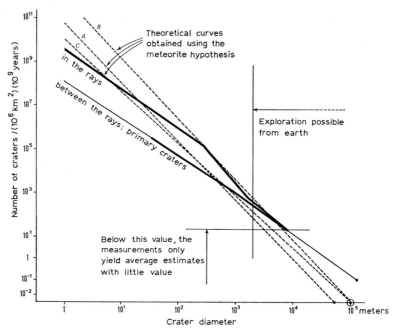

Fig. 30. Relation between the numbers and sizes of observed craters. The solid lines correspond to the measurements.

On any scale, primary craters can be recognized by their often having rays and producing secondaries. Thus by theoretical calculation the observed curve of the total number of craters can be used to obtain the curve for primary craters, which can be determined unambiguously only near the rays of very large craters.

This statistical study argues very strongly in favor of the meteoritic hypothesis. A purely volcanic hypothesis would in fact probably involve a 'threshold': very small volcanic craters probably could not exist. On the other hand, given our present knowledge of meteoroids, we can compute the number of impacts on the moon per unit time for each meteorite size. Several assumptions are necessary. To begin with these involve the velocity with which the meteorites fall on the moon: the relation between the size of the crater and that of the meteorite depends on the impact energy. Another unknown which must be fixed is the relation between the meteorite's size and its impact energy: this energy ($\frac{1}{2} mv^2$) is related to the radius r_m of the meteoroid provided that it is solid (and homogeneous) and of density ρ; for a sphere $m = 4/3\pi\rho r_m^3$; thus r_m is proportional to the $\frac{1}{3}$ power of the energy (case C of Figure 30). On the contrary, if the meteoroid contains hollows, or is inhomogeneous, the relation between m and r_m changes: m may be proportional to a higher power of r_m, if we suppose that large meteoroids are more tightly packed. In Figure 30 the case (A and B) of a law $m \propto r^{3.4}$ is considered. However, it should be remembered that the size of the crater is determined only by the *energy* of the meteorite, and not by its mass, size, or speed separately.

Figure 30 shows that this meteoritic theory gives the right order of magnitude, but predicts not quite enough large craters, and a bit too many small craters. Thus, progress on both the theory and on the statistics remains necessary. There is a hypothesis which might account for the difference between the two curves: first of all, among the large craters, some should be of volcanic, or more precisely internal, origin. Some of the small craters produced by meteorite falls (a larger proportion the smaller the size) might on the other hand be destroyed little by little, covered by a layer of micrometeorites or of lunar 'dust' (produced either by some internal process or by meteorite falls). We shall see below why I have written 'dust'.

According to this hypothesis, secondary craters are produced by the debris thrown up from the lunar surface by the fall of the meteorites producing primary craters. The number of secondary craters n_s produced by a primary crater is estimated from the observed statistics for the large craters ($r > 400$ m) along and between the rays of the well known primary crater Tycho. This number depends on the radius r_p of the primary crater, and we find $n_s \propto r_p^{-s}$, where s is somewhat less than 4. This 'law' may be somewhat arbitrary, since it depends on the extrapolation of a very local set of measurements: does it not depend on the properties (hardness, degree of packing) of the soil where the primary crater was produced? The same can be said for all the aspects of this theory, moreover! Nevertheless by applying this law we can determine the number of primary craters (appearing as a solid line in Figure 30) from the total number observed, and so we can complete the theory with the comparison of two directly related quantities: the number of primary craters of given size, the number of meteorites of given energy...

C. THE NATURE OF LUNAR SOIL

Naturally since recent experiments we know much more about the surface of the moon. Still, it should be noted that, in contrast to the quick conclusions of what may be called the sensational press, the discoveries made by Soviet and U.S. probes, concerning the nature of the moon's surface, confirm the hypotheses that could be formulated from ground-based studies, and (for the time being at least) in no way constitute a revolution on the concepts of astronomers.

A few of the classical results should be stated.

First of all, valuable information is obtained from the general appearance of the craters under varying lighting (according to the moon's phases); the existence of erosion is clearly shown: some craters overlap one another, some of the relief is very soft and rounded; others (such as Copernicus) reveal "sharpened ramparts, bold summits" (Dollfus). The origin of this erosion is clearly related neither to the flow of water nor to that of air, as on earth. From the telescopic examination of the moon, plastic deformation of the rock under local temperature fluctuations as well as the effects of the smallest meteorites have been invoked.

Other aspects of lunar features have been studied: the central peaks of craters, small mountains (of internal origin?) remaining moreover quite rare; rays and rills, straight or irregular cracks, light wrinkles, cliffs... As with the explanation of the vast seas, the hypotheses developed to explain each of the topographical features places a special emphasis on phenomena of internal origin.

The soil proper is also observable. To begin with its color is not uniform. Dark seas, bright haloes surrounding the large craters, the very bright interiors of the youngest craters – all this is clearly related to the reflecting power of the soil. But why these differences? Differences in nature of the soil, or in its age? Polarization studies of the moon show a slight dependence on the point chosen, in particular on its albedo. Lyot and later Dollfus have shown precisely that the polarization of lunar soil varies (as a function of the inclination of the light rays) according to a law (see Chapter VI, pp. 78–83) which "is the same as that for a fine granular deposit of volcanic ash of small dimensions". Recently Dollfus has shown that the residual polarization of lunar soil, in the part illuminated by the earth but not by the sun, depends on the albedo exactly in the same way as the residual polarization of volcanic ash under the same conditions. Figure 31 shows a sampling of volcanic ash having

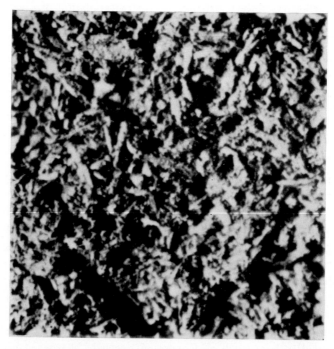

Fig. 31.   The texture of an artificial soil fabricated so as to account for the optical (scattering, polarization) properties of the lunar surface. (Courtesy of A. Dollfus.)

the same polarization properties as the lunar surface. Of course it is not obvious that these materials are the only ones to have such properties; still such ground-based observations showed the likelihood that lunar soil was rough on a small scale, before space research. Certain polarization measurements show that, even on a steep slope, grains of lunar material stick to the surface 'under the influences of forces stronger than weight'.

In addition, optical (infrared) and radio measurements, particularly during eclipses, show that the lunar material is a very poor heat conductor. The coefficient $y = (\kappa \rho C)^{-1/2}$ characterizes heat propagation, (where κ is the thermal conductivity, ρ the density, C the specific heat). The data for the moon yield $y = 1000$. Now for compact terrestrial rocks, $y \sim 20$; for porous rocks (or lava) $y \sim 100$; for divided material (in powder), $y \sim 200$. In order to obtain values of order 1000 in the laboratory, fine powders, measured in a vacuum, are needed.

It is clear that these pre-space-research results are far from justifying the frequent statement that the lunar surface consists of dust. Instead we must decide on what the word 'dust' signifies. On the earth it implies not only that the material is finely divided, but also that it behaves nearly like a liquid. Dust flows, like sand from a dump truck; its surface, like that of liquids in equilibrium, is flat and smooth...

These liquid properties of dust are due to the air contained in the grains of dust and on their surface; the effect of this air, together with the air in between, is to lubricate the dust grains as they slip over one another.

On the contrary, in a vacuum, dust cannot have such liquid properties; no air separates the grains; moreover, if the vacuum persists very long, the air contained within the grains themselves will escape. The solid matter of a dust grain is in direct contact with the neighboring grain or with the adjacent surface; the system 'sticks'. The (so-called Van der Waals) contact forces depend on the reciprocal of the separation distance, to the 7th power. Once this separation is of the same order of magnitude as atomic dimensions, the contact forces are far stronger than gravity forces and the material holds together: in other words the dust grains stick together, leaving empty spaces here and there, and forming a light, strong, irregular material, certainly erodable, and very poorly conducting.

This obviously is independent of the composition of the dust. In any case it should be noted that the roughness, poor conductivity, etc., were known before the direct exploration of the moon, thanks to telescopic measurements and physical and chemical theory.

How then can the difference between the dark and the bright regions of the moon be explained, if the soil is of the same composition? From an examination of the relief, it appears that the bright regions are the younger ones. This means that they are regularly eroded by transport of material due not only to the 'blasts' of meteorite falls on the one hand and to internal dislocative forces on the other but also perhaps to electrostatic forces. The bright regions of relief are soil recently uncovered by erosion. The dark regions of the valleys and plains are regions covered by older aged material, part of which may come from other regions (but from the superficial parts of these other regions). These properties are then easily understood when we note (as laboratory experiments prove) that the bombardment of rocky matter by the protons of the solar corpuscular radiation has the affect of noticeably darkening the target surfaces.

D. SPACE EXPERIMENTS AND THE NATURE OF THE LUNAR SURFACE

To what extent has direct experimentation confirmed or invalidated these views? Photographs of the lunar ground from the surface as taken by the Soviet device which first made a soft landing, or by the Surveyor laboratories, clearly revealed the ex-

Fig. 32.   The texture of lunar soil, as seen from a distance of several tens of centimeters (photograph transmitted by Luna 9 on February 4, 1966).

tremely rough nature (on a scale of a few centimeters) of lunar soil (Figure 32). The sequence of pictures obtained by Surveyor showing a shovel at work digging, shows that it is erodable; the numerous pictures taken by Lunar Orbiter reveal extraordinary details: the slide of a large rock (5 m in diameter) down a lunar slope was captured. Note that pictures taken by Ranger 7 already revealed small rocks of irregular shape: in particular one such, right in the middle of a crater, is clearly the rocky fragment that caused it. Moreover, according to Kuiper, the most likely interpretation of the Ranger observations (which deal mainly with the maria) involves the concept of a 'karst-type' structure of their topography, due to thermal phenomena of internal origin: then the layer of rough dust covering the mare would be very thin.

Exploratory probes do not provide data only on the nature of the surface. They have also confirmed the absence of an atmosphere (less than a millionth of ours), and they have shown that the moon has no significant magnetic field (its magnetic moment is less than 10^{-4} times the earth's magnetic moment).

As we see, our knowledge of the lunar surface advances by successive retouches. The Surveyor experiments will give us the composition, later the geological and petrographic properties will be determined. Undoubtedly exploration by space probes is now the most efficient tool for our knowledge of the moon. However, up to now, rather than contradicting what we thought we knew, such exploration has extended it, filled in some details, and often strikingly confirmed the concepts developed by telescopic research.

E. THE FUTURE OF LUNAR EXPLORATION

How shall we proceed in the future? What questions should be submitted to the exploratory probes? To begin with, the chemical composition of lunar soil is to be determined directly, by way of its scattering properties for α-particles. This experiment has been carried out, and analysis of the data yields a composition similar to that of basaltic rocks. A procedure has been suggested for revealing the presence of water, by sending ahead projectiles from Ranger-type rockets and studying the water vapor ejected by the impact from any hydrated crystals present. Seismological activity also will be studied, and the chemical composition of any atmosphere, no matter how thin, will be determined.

This exploration will remain limited so long as men do not visit the scene. The first manned landing, Apollo XI, took place July 20, 1969. The immediately preceding experiments naturally were essentially aimed at the preparation of this trip: detailed cartography and accurate topography were the mission of the Lunar Orbiters. Besides this, the measurement of the moon's gravitational field, the study of ionic and atomic densities in the lunar environment as well as that of the abundances of high-energy cosmic ray particles and meteoroids of all sizes were all vital elements for this voyage.

Experiments will lead to the establishment of fixed and mobile stations on the moon. Going beyond strictly lunar problems, such stations might include observations analogous to terrestrial observatories (by their programs). (Another volume by the same author, *Les observatoires spatiaux*, will deal with these.) Certainly the lower gravity and the absence of an atmosphere will facilitate the construction of enormous devices, and both astronomy and astrophysics will develop rapidly with such stations, manned or automatic. There remains a question which is still difficult to answer: won't the stability of these stations be affected by tremors of the lunar surface, whether these be due to seismological activity or to the fall of heavy meteorites on the moon?

It should be clear that such exploits are not attempted simply for the sake of the elegance of the act or the beauty of an excellent picture. In any case, that is not the goal of the astronomers who take part in the organization of space experiments. It is a question of methodically searching for the answers to the problems posed by the gaps in all our previous knowledge. In this field as in all other fields of astronomy, space techniques are the powerful companions of classical techniques, and not their victorious rivals.

A recent example confirms this. While correcting the proofs of this book, new data appear. In particular, the study of the motion of circumlunar satellites has revealed marked density irregularities: under the *maria* ultra-dense mass concentrations – so-called 'mascons' – exist, perhaps gigantic meteorites buried beneath the surface. This discovery should lead to a revision of some traditional astronomical notions, in particular that of 'ephemerid time' which is closely linked to astronomical observations of the moon's motion.

2. The Exploration of Mars

Mars is known to have a very thin atmosphere. Temperatures measured there are of the same order of magnitude as temperatures on the earth. Spectroscopy reveals an atmosphere rich in carbon dioxide (CO_2), poorer in oxygen (O_2) and water vapor, with traces of other gases ($N_2O...$). Its surface (whose colors and polarization have been studied) is no doubt mostly dusty, and may consist of materials similar to limonite ($2Fe_2O_3, 3H_2O$). The white polar caps, consisting of a frost of fine spherules, expand in winter and shrink in summer.

Color changes of the surface are striking; the dark green regions turn yellowish in the Martian autumn, and the existence of plant life (thallophytes, bryophytes, algae?) has been suggested.

Research by space probes has modified these telescopic conclusions only slightly.

However, it has revealed new fundamental facts (see Table II, pp. 9–11): the very low magnetism of Mars, the nature of the surface, dotted with craters similar in size and appearance to lunar craters, are important additional pieces of information. Thus, the nearly total absence of a magnetic field ($\frac{1}{3000}$ the magnetic moment of the earth) is a valuable clue regarding the conditions of life on Mars, subjected to intense corpuscular bombardment. This same fact must be explained in the framework of cosmogonic theory of the origin of the solar system. The existence of craters confirms the meteoritic origin of these features: Mars and the moon have in common the thinness or absence of an atmosphere and the absence of erosion. Mars now resembles a larger moon! (see Figure 33).

Let us wait for a more thorough exploration: at that time, data on possible life on Mars will certainly be among the most important pieces of information that can be obtained.

3. The Exploration of Venus

The passage of several space probes in the neighborhood of Venus, as well as the landing of a device on the planet, have also helped to elucidate a number of Venusian characteristics.

Of course, it was already known that the cloudy atmosphere consisted essentially

Fig. 33. A Martian crater as seen from 10000 km above the surface of Mars (NASA photograph, Mariner 4, July 14, 1965). The dimensions of this crater ($\sim$ 200 km in diameter) are comparable to those of lunar craters. Several smaller craters are also visible on this photograph.

of CO_2. However, fairly accurate measurements of temperature and magnetism have now been carried out. The magnetic field is that of a dipole whose moment is 10–30 times weaker than that of the magnet 'earth'. The temperature is that of a 'hothouse'. It rises from 92 °C at the base of the cloud layer to 260° near the ground (some measurements go even higher: 426 °C at the ground). Thus no trace of liquid water is to be found there: any oceans will have been vaporized and so exist in the clouds... Still these data remain preliminary, and one can hardly say much more: the 10 years to come will be a decade of planetary exploration (if budgets allow!) and we shall learn much about the topography, soil, and atmosphere of Venus that remains hidden from our telescopes by the clouds. Here again we must work patiently, wait and see.

4. Meteoroidal Particles and Dust

The study of the meteoroids of the interplanetary dust – as well as that of the elementary particles: Protons, neutrons, mesons..., which constitute the earth's environment, is well advanced.* As for further exploration... it is a bit early to write a separate paragraph on this, and we were after all right to discuss it within a general survey. But let us wait several years: then no doubt we will be able to arrive at specific ideas concerning the genesis of the planets and the origin of the solar system and of the sun itself, as well as the interaction of the solar system with the interstellar medium which fills the galaxy, using the data brought back by such exploration.**

* See *Space observatories*, by the same author.
** Since this chapter has been written, the success of Apollo XII, the perilous trip of Apollo XIII, scientifically a failure, the analysis of moon fragments brought back by the Apollo XI and XII missions, the results from the Mariner missions around Mars – all have brought to light many new facts of great interest. They do not contradict what has been written in this chapter, but they make it somewhat obsolete. At this stage it was difficult to modify this chapter deeply, but we certainly can predict a blooming of publications on the new conquests.

CHAPTER VI

THE PLURALITY OF INHABITED WORLDS

1. With Fontenelle and his Marquise

We have no intention here of plunging into literature, or science fiction, or even scientific speculation. Still, can we completely ignore what has been written?

It is quite clear that man has always dreamt of living beings beyond his ken, situated either in the infinitesimally small or in unexplored worlds of the infinite. Cosmic dreams abound in literature. Sometimes this has been the pretext for some amusing philosophy; in Micromégas, Voltaire expounded his point of view through Sirius... Cyrano de Bergerac and Swift also had things to prove... After all, their selenites and Laputians are but men – terrestrials – in disguise.

Fontenelle went into the problem more seriously. In his delightful dialogues (on the 'plurality of inhabited worlds') with an amiable marquise, it is clear that life here, and there, cannot be of the same character. If I may be allowed to quote once more:

Why should there not be selenites?

"This is why I have come to believe in the inhabitants of the Moon. Suppose that there had never been any commerce between Paris & Saint-Denis, and that a Citizen of Paris who has never left his City, should climb the Towers of Notre-Dame, & see Saint-Denis in the distance; let us ask him if he believes that Saint-Denis is inhabited, like Paris. He will answer no straight off; for, he will say, I can see the inhabitants of Paris, but as for those of Saint-Denis, I can't see them at all, and no one has ever heard of them. Someone will argue with him, saying that while it is true that one cannot see the inhabitants of Saint-Denis from the Towers of Notre-Dame, it is because of the distance; everything that one can see of Saint-Denis looks very much like Paris; Saint-Denis has belfries, houses, and walls, & might resemble Paris all the more in being inhabited. All that will not succeed in convincing our Citizen, he will always persist in affirming that Saint-Denis has no inhabitants, since he can see no one there. Now our Saint-Denis is the Moon, & each of us is this Citizen of Paris who has never left his City."

"Ah!", interrupts the Marquise, "you are not fair with us, we are not at all so stupid as your Citizen; inasmuch as he sees that Saint-Denis is just like Paris, he must be out of his wits to believe it uninhabited; but the Moon is not at all like the Earth." "Take care, Madame", I answer; "for if it turns out that the Moon resembles the Earth in every way, you will be obliged to believe that the Moon is inhabited." "I admit", she answers, "that there is no way to avoid that, & I see that you have a confident air that already frightens me."

Let us then also be confident. And, like Fontenelle, let us leave for the moon, first with the scientists, then with Arioste:

"One place is called Copernicus, another Archimedes, a third Galileo; there is a Cape of Dreams, a Sea of Rains, a Sea of Nectar, a Sea of Crises; to sum up, the description of the Moon is so precise, that a scientist up there at this moment would no more get lost than I should in Paris."

"But", she replies, "I should like to know still more of the details of the lay of the land." "It is impossible", I say, "for these gentlemen of the Observatory to tell you more; you must ask this of

Astolfe, who was taken to the Moon by Saint John. I am referring to one of the most pleasant *folies* of Arioste, & I am sure that you will be pleased to hear it...

... This is what it is about. Roland the nephew of Charlemagne went mad, because the beautiful Angélique preferred Médor to him. One day Astolfe, a fine Paladin, was in the Terrestrial Paradise, at the peak of a very high mountain, where his Hippogryph had carried him. There he met Saint John, who told him that in order to cure Roland's madness, they would have to travel to the Moon together. Astolfe, who was eager to see the world, agreed right away, & immediately there appeared a chariot of fire to fly away with the Apostle & the Paladin. Now since Astolfe was not much of a Philosopher, he was quite surprised to see the Moon much larger than it had appeared to him from the Earth. He was even more surprised to see other rivers, other lakes, other mountains, cities, forests, &, what would have surprised me quite a bit too, Nymphs hunting in these forests. But the most extraordinary thing he saw on the Moon was a valley in which everything of every sort that was lost on Earth was to be found."

Inhabitants? Therefore air! But what air?

"I am again overjoyed that you have given inhabitants to the Moon. I am still happier that you have given it an Air specially surrounding it; it seems to me now that without that a Planet would be too naked."

"These two different Airs, I continued, help to prevent communication between the two Planets. If we only had to fly, how do we know, as I was saying to you yesterday, that we shall not be able to fly very well some day? I admit however that this appears unlikely. The great distance of the Moon from the Earth would be still another difficulty to be overcome, and certainly a considerable one; but even if this difficulty did not exist, even if the two Planets were very close to one another, it would not be possible to go from the Air of the one to the Air of the other. Water is the Air of Fish, and they never enter the Air of the Birds, nor do the Birds enter the Air of the Fish; it is not the distance that stops them, it is that each is imprisoned by the Air he breathes. We find that ours is mixed with thicker & grosser vapors than that of the Moon. As a result an Inhabitant of the Moon arriving at the limits of our World would drown upon entering our Air, & we would see him fall dead on the Earth."

And in this air, what inhabitants?

"Certainly we would not find men there."

"What sort of people would they be then?", asked the Marquise impatiently.

"In all good faith, Madame", I answered, "I have no idea. If it were possible that we should be endowed with reason & still not be men, & if moreover we inhabited the Moon, how would we imagine the existence below on Earth of this bizarre species of creatures called Humanity? Could we picture a creature with such mad passions, & such wise reflections; so short a life, & so long an outlook; so much science on nearly useless things, & so much ignorance of the most important; such a love of liberty, & such an aptitude for slavery; such a strong desire for happiness, & so great an incapacity for it? The people of the Moon would need to have quite sharp minds, to guess all that. We see ourselves incessantly, & we still have to guess about our own nature. The only explanation one has found has been to say that the Gods were drunk with nectar when they made man, & when later they looked upon their work with a cool head, they could not stop themselves from laughing."

"So we are quite safe from the people of the Moon", said the Marquise, "they will not guess our existence; but I would be happy if we could learn of them; for in truth it is disquieting to know that they are up there on the Moon that we see, & that we cannot imagine how they are made."

"Why", I answered, "aren't you worried about the inhabitants of the great Austral Land which is still entirely unknown to us? Both we & they are transported by the same vessel, they at the bow, & we at the stern. You see there is no communication from bow to stern, & at one end of the boat we have no idea who is at the other end, or what they are doing; & you would like to know what is happening on the Moon, that other vessel floating far from us in the Heavens?"

Of course, difficult to picture... And what about the bees after all! There we have some quite strange beings, living among us:

"... Then I explained to her the Natural History of the Bees, of which she knew just the name. From which you can see", I continued, "that simply in transporting to other Planets things taking place on ours, we would imagine oddities which might appear extravagant, and still be very real, & we would imagine an infinite number of them; & Madame, you should realize that the History of the Insects is filled with them." "I can believe that easily", she answered. "To take just the example of Silkworms, which I know better than the Bees, they would supply us with rather surprising Peoples, going through a metamorphosis & so changing completely, crawling during one part of their lives, & flying during the other; & how can I say? a hundred thousand other marvels – the different characters & customs of all these unknown Inhabitants. My imagination is working on the plane you have suggested, & I can even go so far as to make out their faces. I could not describe them to you, but still I see something."

"As for those faces", I answered, "may I suggest that you leave it up to the Dreams that you will have tonight. Tomorrow we shall see if they have taught you something of the Inhabitants of some Planet."

In any case, it is not absurd to differ from the accepted notions; in this field many hypotheses are allowed. What is true on one side of the Pyrénées may be false on the other side... Or is it the Urals, or the Great Wall:

"I have just told you", I answered, "all the news I have of the Heavens, & I do not think there are any more recent. I am annoyed that they are not so surprising & so marvelous as some observations I was reading the other day in a summary of the Annals of China, written in Latin. There we see a thousand stars at once falling from the Sky into the sea with a great fracas, or else dissolving & disappearing in rain. That was not observed only once in China; I have found the observation at two rather distant epochs, not counting a star exploding noisily like a rocket in the West. It is unfortunate that such spectacles be reserved for the Chinese, & our countries have never had their share. It was not long ago that all our Philosophers thought that experience supported their contention that the Heavens & all the celestial bodies were incorruptible & unchangeable, & at the same time other men at the other end of the Earth saw stars dissolving by the thousands, which is rather different."

"But", she said, "haven't I always heard that the Chinese were very great Astronomers?"
"That is true", I replied; "but..."

Let us leave behind Fontenelle, his marquise, and his hardly extravagant imagination. Our contemporaries are not all so measured. Often they tend to fantastic imagination rather than rational deduction. Wells produces Martians which are less peaceable but just as unlikely as the Sirians of Voltaire, and Hoyle's black cloud is more than anything else a brilliant and amusing device to make fun of our governments.

We will not speak here of the avalanche of flying saucers and Martians unloaded *en masse* by the so-called sensational press (and even the other press), always invisible to the majority of men and to the totality of scientists...

Of course, it might seem that this problem should be eliminated *a priori* in a book whose subject is 'experimental astronomy'. However, we would not like to limit the definition of 'space research' too narrowly. To the extent that the possibility of communication with extraterrestrial worlds depends in the first place on astronautics and in the second place on radioastronomy, we shall consider that the problem of extraterrestrial life remains within the scope of this book.

However, we shall say right away that very little is known about life in the universe. We shall go through the inventory quickly, and the examination of our hopes for further knowledge in this field will be quite rapid too. No doubt we must make an appointment for the end of the century. In 1999 we shall certainly know much more. But let us take a closer look.

2. Life and the Biosphere

Is there life on the other planets or satellites of the solar system? Certainly the very fact that we are human and earth-bound introduces an element of anthropomorphic subjectivity directly into the methods by which we search for and might discover extraterrestrial life. We can conceive of life taking very different forms from those on earth; here carbon biochemistry reigns. However, is a biochemistry based e.g. on phosphorus, inconceivable? So it is said, although we do not know. In any case, it should be expected that the first species of life we discover will resemble those with which we are familiar on earth: as for the others, we will at first most probably have great difficulty in recognizing them as forms of life! Can we pick them out by their movement? Crystallizations and turbulent motions exist in the field of classical physical chemistry: Reproduction? It took a long time for the fission of protozoans to be recognized as such; Nutrition, respiration? All this is very vague, and it seems there is a continuum of increasing complexity going from the most inert matter to the most highly developed living being... What then is life, in a new set of physico-chemical conditions? Having asked this question, we now can only (and even should only) limit ourselves to an anthropomorphic search for extraterrestrial life, life similar to that on earth...

Given this unfortunate but necessary restriction, which we shall adopt once and for all here, i.e.: that we shall deal with life analogous to that on earth, where can we first expect to find such life?

Even on earth, it is not easy to define the conditions under which life can continue. In any case, we cannot relate the investigations of the conditions necessary for the 'appearance' of life to the study of present life in the universe, for the past history of celestial bodies is so rich, and naturally poorly known, that it is likely that millions of objects have been, are or will be in a state favorable to the appearance of life, even if the necessary conditions are extremely restrictive. Even on earth, life can 'continue' under conditions which do not cover the entire globe, but which nonetheless remain rather broad. Using the term introduced by Lamarck, we shall call the region in which these conditions are satisfied, the *biosphere*.

From the ground, the biosphere extends upward to an altitude of about 10000 m; bacteria and spores without chlorophyll, as well as birds, have been observed at 8000 m; certain species of spiders have been found on Mount Everest (at 7000 m). The oceans are included in the biosphere. Under the continents life extends to depths of 1 or 2 km; bacteria in ground water, in oil... As for the green plants, certain species exist above 6000 m on Mt. Everest; algae grow as deep as 300 m below sea level. Moreover this extension is most of all limited by the food supply in organic material of vegetable origin. For Vernadsky the biosphere is the region in which water can exist in the liquid state, numerous phase transitions (liquid-solid, solid-gaseous, gaseous-liquid...) exist, and radiant energy (with the exception of short wavelengths, which are in fact destructive) is absorbed by partially opaque matter. This last point is fundamental, for the green plants play the role of intermediary in transforming the radiant energy of the sun into the chemical energy of various molecules. One of the fundamental reactions of photosynthesis is assimilation by chlorophyll, consisting essentially in the production of carbohydrates from carbon dioxide and water. This reaction reduces to:

$$6 \, CO_2 + 6 \, H_2O \rightarrow C_6H_{12}O_6 + 6 \, O_2 \, .$$

Of course, in reality the reactions are more complicated and pass through numerous intermediate steps, so that glucose is certainly not the primary product.

Obviously the conditions defined above exclude the maintenance of life in a major part of the universe: where can conditions be close to what they are in the biosphere?

In the first place the stars are excluded: the temperature there is too high for anything but gases to exist; even the simplest molecules (e.g. diatomic) can be found only in the coolest stars.

The interstellar medium, far from stars, must also be excluded, at least for the time being. It would seem that the radiation field there must be too weak to sustain the slightest amount of photosynthesis. Also, contrary to the old theories of Arrhenius, it is clear now that the interplanetary medium must also be excluded because of short wavelength radiation, which could destroy any life there. Only under exceptional circumstances could life appear in or rather pass through the interplanetary medium during some fairly rapid migration, but we should not expect to encounter life in normal development there.

Thus, only objects close to stars, where the above conditions might be satisfied, are left: these are essentially the *planets*.

In any case we can be sure of one thing. The planetary phenomenon is an extremely common phenomenon in the universe. Ten major planets and thousands of planetoids turn about the sun. And already many years ago, Van de Kamp showed that a good third of the stars near the sun were almost certainly accompanied by a significant planetary system.

Before we examine whether life might be found on such planets, and what the position of astronomy is on that, let us in any case see what the situation is on the planets of our solar system.

3. The Solar System

Here again we must begin by elimination; bodies which are too small cannot conserve an atmosphere: this is the situation of the moon and the asteroids. Therefore it would seem likely that chlorophyllic assimilation cannot take place there; however, can the presence of anaerobic organisms be excluded? Since life is possible in oil and underground water on earth, could it not exist in lunar rocks? We are no doubt close to the answer. For the time being, however, we must consider that the possibility of life under such conditions is quite low.

The major planets? In any case, Mercury, like the moon, has no atmosphere. Jupiter, Saturn, Uranus and Neptune are probably too cold. Pluto is still colder. The earth (of course!...), Venus and Mars are left.

We cannot yet say very much about Venus. Table III gives an idea of the temperatures and chemical composition of its atmosphere: certainly there is nothing there that would *a priori* force us to consider life impossible on Venus. However, this is all that we know today (1967) and neither Mariner nor Zond has given us any more information in this respect. Let us wait!

In any case we shall quote one opinion, Sagan's. Given the high temperature of Venus, he concludes that its surface must be desert-like: precipitation cannot reach the ground, and all the water remains at high altitude in clouds. This argument leads to the conclusion that Venusian life is unlikely. However, it remains quite speculative.

We are left with the only planet outside of earth, where some observations lead us to suspect the existence of life. The specialty of 'astrobiology' has been formed around observations of Mars, inspired by the Soviet astronomer Tykhov. What can we say about it?

There are quite a few astrophysical methods for observing Mars. Since its apparent diameter varies from 4″ of arc to 26″.6 under the best conditions, many details of the surface can be observed visually. Color and appearance vary both from place to place and with time. Fairly good spectra can be obtained of the features; their polarization and diffusive properties can also be measured, at various wavelengths and under varying angles of incidence of the sun's rays illuminating the Martian atmosphere.

The study of the Martian atmosphere reveals first of all that it is very thin, poor in oxygen but rich in carbon dioxide. Water vapor does exist, but in small quantities (as we have seen in the preceding chapter): a new and recent proof of this was given by the discovery of craters implying the absence of erosion.

We can compare the columns corresponding to Mars and the earth in Table III. It is clear from this comparison that any life on Mars could not closely resemble terrestrial life. Devices called 'phytotrons' are available, in which the conditions of the Martian atmosphere (low pressure, high abundance of carbon dioxide, low humidity, absorption of blue radiation...) with respect to life can be reconstituted. Experiments show that mosses, lichens, and certain bacteria could survive and develop there. However, the limitations of this experiment must be noted: which organisms could live within the Martian rocks, or in any underground gases or liquids that might be present there? Also the experiment itself is anthropomorphic to the extent that it takes only certain parameters into account, and these parameters may perhaps be important only for those organisms with which we are most familiar. Are they the only important ones, moreover? Might there not be some very special forms of life in the presence of a strong magnetic field, for example? At least it would seem that the differences between the earth and Mars have been taken into account objectively: such instruments are no doubt objective tools when used as instruments for the differential study of life on the earth and on Mars.

What then does the direct observation of Mars tell us at present?

First of all we shall mention the anthropomorphic Martian canals of Schiaparelli and Lowell, but we shall immediately reject them in the bag filled with fairy fables.

TABLE III

Chemical composition and temperatures of planetary atmospheres

	Mercury	Venus	Earth	Mars	Jupiter	Saturn	Uranus	Neptune	Titan (satellite of Saturn)
A	+[a]	–	–	4	–	–	–	–	–
H_2	–	–	–	–	270×10^3	40×10^3	135×10^3	150×10^3	–
O_2	–	+	1.68×10^3	N[b]	–	–	–	–	–
N_2	–	–	6.25×10^3	300	–	–	–	–	–
He	–	–	–	–	–	+	370×10^3	370×10^3	–
CO_2	–	Abundant, 10^3 above clouds	2.20	50	–	–	–	–	–
H_2O	–	~70 μm (precipitable)	Variable	8–20 μm (precipitable)	–	–	–	–	–
CO	–	N	–	N	–	–	–	–	–
N_2O	–	N	0.008	–	–	–	–	–	–
CH_4	–	N	0.017	N	150	350	3×10^3	5×10^3	250
NH_3	–	N	–	N	7	2	–	–	N
SO_2	–	–	–	N	–	–	–	–	–
O_3	–	–	0.003	N	–	–	–	–	–
Temperature:									
Theoretical max.	625	324	349	306	131	95	67	53	119
mean	–	229	246	216	93	68	47	38	84
Observed max.	613	–	314	287	130–165 (G)	–	–	–	–
mean	–	700 (G)[c] (near the surface)	–	210	–	93	–	–	–

[a] + denotes non-quantitative evidence.

[b] N denotes a negative test yielding only an upper limit.

[c] G indicates that there is a strong 'greenhouse' heating effect due to atmospheric opacity!

Note: The table entries are given in m/atm, i.e. the thickness in meters of the atmosphere, supposed reduced to 'standard' conditions of temperature and pressure.

In fact the telescopic appearance of Mars leads one to distinguish the white polar caps and dark spots of varying size, against a light yellow-orange background. There is a significant variation in the appearance of these details; the comparison of these with the seasonal variations of terrestrial vegetation suggests the possible existence of extensive plant life. As early as 1927, Slipher considered that the dark blue-green regions might be due to vegetation.

These seasonal variations can be described as on earth, with the two hemispheres naturally a half-cycle out of phase. We shall recall that the inclination of the equatorial plane of rotation of Mars to its orbital plane is 24° (compared to 23°27′ for the earth), and that its period of revolution is 687 (365 for the earth). Thus the evolution of the surface of Mars can be described as follows: At the beginning of spring, the spots close to the pole become sharper. Their contrast increases; in the meantime the polar cap recedes and shrinks considerably, apparently melting or sublimating, leaving behind bright isolated points, probably due to the Martian relief which they thus reveal. A wave of change seems to start from the polar cap: the dark regions at low latitudes seem to become darker only at the end of spring, after the phenomenon has already begun to weaken at higher latitudes. In winter the 'dark' spots appear quite pale, before the wave of darkening moves over them on its way from the pole to the equator. It is possible that variations in humidity (transported by some meridional circulation) are the cause of these color changes. But it should be noted that as early as 1930 Antoniadi also observed certain dark spots turning brown and red in the Martian autumn, strangely similar to autumn on earth.

These seasonal variations are difficult to observe; they are complicated and partly camouflaged by other variable phenomena. As we know, the Martian atmosphere is dotted with white clouds (fog, haze perhaps due to ice crystals) or swept by dust storms (winds of 'Martian sand' whose yellow-orange color may be due to iron oxides), and sometimes covered by a blue haze (morning fogs due to water droplets, perhaps analogous to noctiluscent clouds on earth). Moreover, there are irregular phenomena taking place on the surface, changing the appearance of certain Martian regions from year to year. Sometimes these changes occur over a period of several years.

Polarization measurements furnish us with more information on the nature of these variations of the Martian surface.

The polarization depends on the phase angle, and depends strongly on the nature of the light-diffusing medium, in this case the nature of the Martian surface. The degree of polarization

$$p = \frac{(I_1 - I_2)}{(I_1 + I_2)} = f(V)$$

is measured, where I_1 and I_2 are the amplitudes of the intensity (electric vector) in the plane of 'vision' (Figure 34) and perpendicular to this plane. Note that the relation between p and V depends only slightly on the other parameters (although in principle the orientation of the normal to the diffusing surface with respect to the plane of vision should also enter into account).

The angle V can go from 0 to 45° in the case of Mars; Figure 35 shows the polarization as a function of V. This type of measurement, first carried out by Lyot in 1922, was elaborated by Dollfus in the years 1948–56. Following this author, measurements of the orange regions (Martian deserts near the center of the disk), of the dark regions and of the polar caps, are plotted separately. In order to bring out the degree of 'individuality' of such curves, we have plotted the analogous curves concerning the moon and Venus on the same figure, as well as measurements obtained from various terrestrial materials, on Figures 36 and 37.

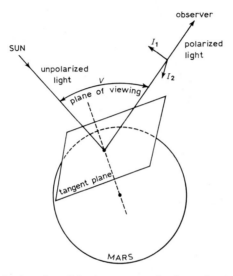

Fig. 34.   Polarization of sunlight due to scattering by a planetary atmosphere.

It should be noted (Figure 35) that the seasonal variations are accompanied by significant variations in polarization.

Of course, clouds also produce a parasite polarization which is not always easily eliminated. However, when we can do this, how can we interpret the measured polarization values?

The answer is clear for the yellow deserts. The curve for limonite is in perfect agreement; the Martian deserts are probably composed of limonite (or of a very similar mineral) in pulverized form.

For the polar regions, laboratory studies have shown that frosts made of fine needles of ice give a satisfactory explanation. This frost would sublimate under solar radiation.

However, the problem of the dark regions is not at all so simple! Terrestrial minerals have given clearly negative results. The study of many phanerogams and thalloid cryptogams has been negative. On the other hand, it seems that the microcryptogams have properties similar to those of the dark Martian regions. Dollfus argues that the coloring of some of these micro-organisms could go through variations similar to those of the Martian surface: this is true for monocellular algae, one variety of which

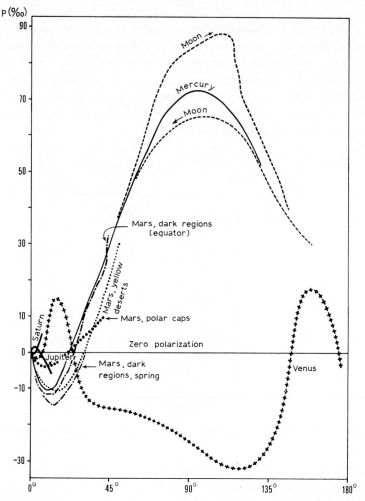

Fig. 35.   Polarization by the atmospheres or the surfaces of the planets and the moon. The angle V, defined in Figure 34, is the abscissa here. The ordinate is the proportion of polarized light, in thousandths.

(*Nivallis*) grows in high mountain regions and colors the snow red. Certain chromogenic bacteria, such as *Bacillus pyocyaneus* which are blue-violet, have very sensitive coloring effects. While these comparisons remain quite preliminary, they are in any case extremely suggestive: for the time being they are the basic evidence, the 'proof' (though it can be seen that one should not be too positive about this) of the existence of life outside of the earth.

Another in fact very serious argument arises from spectrographic analysis.

Already a long time ago Slipher showed that the dark regions of Mars do not have the reflection spectrum of chlorophyll; this was confirmed by Sharonov and Kuiper. However, Tykhov has shown that under the extreme conditions of a very dry climate,

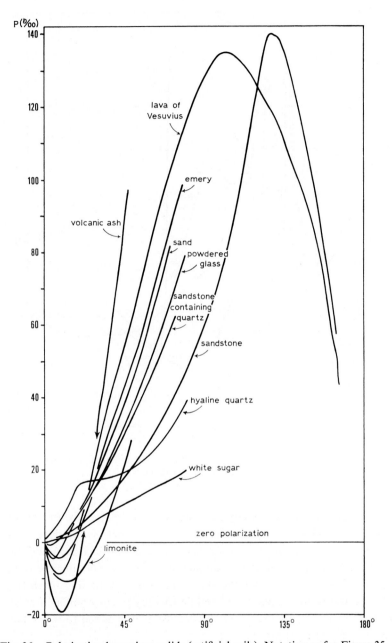

Fig. 36. Polarization by various solids (artificial soils). Notation as for Figure 35.

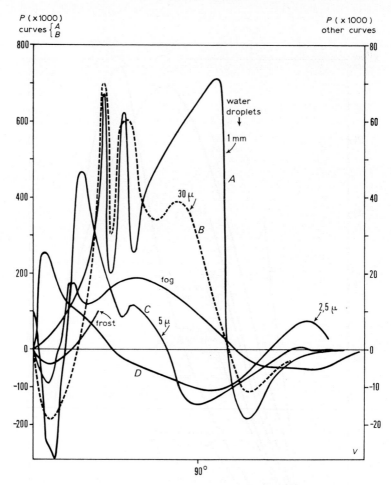

Fig. 37.   Polarization by various fogs, by water droplets of different diameters, by frost. Same
notation as Figure 35.

the reflection spectrum of chlorophyll can weaken. More recently Sinton was able to obtain infrared spectra of the dark regions of the planet. These dark regions have absorption spectrum presenting three molecular bands at 3.43, 3.56 and 3.67 μ. The first two are characteristic of many complex molecules containing the radical CH_2, both organic and inorganic. But the third one seems to be due to carbohydrates. Could this be the result of plant life? While this argument remains highly disputed today and in any case is by itself quite inconclusive, it is added to the others to form the basis of a strong presumption.

   We can best summarize this point by quoting Harold Urey:

No suggestion has ever been made of any inorganic substance likely to be present on an arid and lifeless planet that would supply the observed colors [and one might also say the polarization – J.C.P.] and their seasonal variations.

Elsewhere, he shows how fundamental this question is, and although we have no definite answer yet, it seems highly probable the answer will be: 'Life!'

The positive proof that life in any form depending on the remarkable chemistry of carbon compounds exists on Mars would give us one of the most fascinating facts of all science.

Apart from Mars (and of course the earth), it seems premature to consider at present the possible existence of life in the solar system. However, we would like to refer to the discussion which accompanied the physico-chemical study of certain fragments of large meteorites, in particular the Orgueil fall of May 14, 1864, which was recently re-examined by several research teams. Unfortunately the press (a certain part of it!) in jumping on this affair seriously distorted the philosophy of the research. Some of the investigators (in particular Claus and Nagy) believe they have found organic fragments (which do not necessarily imply life, but in any case do suggest it very strongly) in these meteorites. On the other hand other investigators (Anders, Deflandre) believe that the organic particles found are the results of terrestrial contamination of the meteorite fragments, invalidating the experimental results. Without new and highly systematic research, no conclusion seems possible at present.

4. Other Planetary Systems in the Universe

Now while we can perform direct chemical analyses of meteorites, and analyze the radiation of the major planets of the solar system, we have only extremely tenuous information concerning extra-solar planetary systems. Even the light of a planet ten times larger than our gigantic Jupiter could not be distinguished next to the light of its star – and this is so even for the star closest to the sun, and even if the planet were ten times more distant than Jupiter from its star: even such a planetary system would be much too distant from us to be observed directly!

On the other hand its mass can be determined indirectly, since its presence displaces the center of gravity of the planetary system away from the center of the star. Thus the star will appear to move around this barycenter, and using the orbit which is observable (and measurable with respect to the fixed field stars) the mass of the totality of the perturbing planets, reduced to the barycenter of these planets, can be determined.

We shall make a simple calculation. Let us consider a solar-type star E of mass M and of radius R, located at a distance $D=1$ parsec from the sun, accompanied by a planet of mass $m=1/50$ and radius $r=R/10$ (and thus much denser than the star itself, which is perfectly normal) revolving about E at a distance $a=10$ AU. What would we observe?

Kepler's third law gives us the period of revolution T through the equation:

$$\frac{a^3}{T^2} = \frac{G(M+m)}{4\pi^2}, \tag{1}$$

from which we obtain:

$$T = \frac{2\pi a^{3/2}}{\sqrt{G(M+m)}} \sim \frac{2\pi a^{3/2}}{\sqrt{GM}}. \tag{2}$$

Numerically this gives $T \sim 3.8 \times 10^8$ sec ~ 10 years.

The period would be 5 months for $a = 1$ AU. Thus these periods are easily observed, comparable not only to those of the planets of our solar system, but also to the periods of perfectly standard visual double stars.

The distance separating the star from its barycenter is easily computed; we have:

$$Md = m(a - d), \tag{3}$$

and so

$$d = a \frac{m}{M+m} = \frac{1}{50} a. \tag{4}$$

The corresponding apparent angular separation is then:

$$\delta = \frac{d}{D} = \frac{1}{50} \frac{a}{D} \sim \frac{a}{50}, \tag{5}$$

where a is in AU and δ is in sec of arc. Consequently,

for $a = 10$ AU, $\delta \sim 0.200''$;
for $a = \ 1$ AU, $\delta \sim 0.020''$.

An apparent displacement of this amplitude is perfectly measurable on astrometric photographs of the Carte du Ciel type. The principal difficulties arise from the fact that on such plates the scattering of the starlight in the photographic emulsion produces an image a few microns in diameter, which corresponds at best to a second of arc rather than to a hundredth of a second of arc. Thus one must measure carefully the position of the 'center of gravity' of the darkened spot on the plate which constitutes the image of the star. This procedure only makes sense if this spot is circular. This is not always (and not even often) the case. To carry out such measurements Van de Kamp had to develop techniques of high accuracy in years of efforts. Astrometric reflecting telescopes are the instruments best suited for this sort of measurements, and one – the only one of its kind today – has recently been installed at Flagstaff, Ariz. (U.S.A.).

We can check that even such a large planet will be invisible; let us suppose its albedo equal to unity (i.e. it is a perfect reflector, which is the most favorable case for detection). The planet intercepts a fraction $\frac{1}{4}(r/a)^2$ (Figure 38) of the flux F of stellar radiation, and one half of its visible disk radiates toward us. We observe a certain fraction of this reflected light, and this fraction is the same as the fraction we observe of the directly radiated stellar light. Therefore, the difference in magnitude between the planet and the star will be of order:

$$\Delta m = 2.5 \log\left(\frac{1}{4}\frac{r^2}{a^2}\right) \sim -23 \text{ magnitudes} \tag{6}$$

if $a = 10$ AU, and -18 magnitudes if $a = 1$ AU.

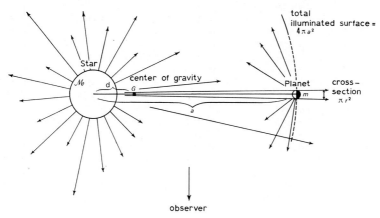

Fig. 38.   Visibility of a stellar planet.

This difference is far too large (the star being of magnitude 7 in this case) for the planet to be detectable on the plate. In the two cases treated here, the apparent separation of the planet and the star would be $10''$ ($a = 10$ AU) and $1''$ ($a = 1$ AU). There would be no trouble from superposition of the images of the star and the planet, but the planet would be far too faint! It can easily be shown that this problem is still far more difficult than the search for the faintest satellites of planets of the solar system!

Despite these problems, Van de Kamp was able to announce the (probable) detection of several planetary systems, in 1948; Table IV gives the principal characteristics of these systems, as well as those of several stars studied since then.

Since 1948, only a small number of publications have been devoted to this question. In fact, however, this is quite normal, when one considers the periods of revolution of planetary systems: nearby stars must be followed for years in order to detect this sort of perturbation, and the difficulty increases with the distance of the star. This results in a rather odd selection effect in the above table. If we take the threshold of detectability of the orbital motion to be of the order of $0''.002$, we have:

$$\delta = \frac{1}{50}\frac{a}{D} = 0''.002 \simeq 10^{-8} \text{ radian}, \tag{7}$$

so that at the limit of detectability the relation between a and D is:

$$a = 5 \times 10^{-7}D \, (a \text{ and } D \text{ in the same units}). \tag{8}$$

TABLE IV

List of stars known to have invisible companions

No.	Name	π_{trig}	P	a	$M/M_\odot$
1	Proxima Centauri	0″.762	2.47	0″.010	0.0018
2	Barnard's star	0″.530	24	0″.0245	0.0015
3	BD + 36°2147 = Lalande 21185	0″.398	8.0	0″.0336	0.01
4	61 Cygni	0″.292	4.8	0″.0102	0.008
5	Krüger 60A	0″.249	16	$\alpha = 0″.03$	0.009…0.025
6	BD + 20°2465	0″.213	26.5	0″.11	0.032
7	o^2 Eridani	0″.201	2.99	0″.012	0.029
8	70 Ophiuchi	0″.193	17	0″.015	0.008…0.012
			9.89	0″.014	0.012…0.015
9	Ci 2354	0″.184	10.8	0″.028	$\geqslant 0.02$
10	ξ Bootis	0″.145	2.2	0″.02	0.1
11	υ Herculis	0″.117	8 or 16 y	0″.5 or 0″.15	–
12	Ross 434	0″.064	1.3	0″.038	–
13	δ Aquilae	0″.062	3.4	0″.05	0.5…0.8
14	BD + 11°2625	0″.061	12.4	0″.15	–
15	ξ Aquariae	0″.040	25	0″.08	0.6
16	16 Cygni A⎱	0″.039	⎰ 1.61	0″.012	–
17	16 Cygni B⎰		⎱ 1.59	0″.031	–
18	ξ Cancri C	0″.039	17.5	0″.191	0.9
19	μ Draconis	0″.033	3.2	0″.026	0.6

π_{trig} = trigonometric parallax.
a = apparent semi-major axis of the orbit of the perturbed principal component.
α = apparent amplitude in right ascension.
P = period of revolution in years.
M = mass of the companion ($M_{\text{Jupiter}} \sim 0.001\ M_\odot$).

This is very close to the relation that can be found between columns 3 and 5 of Table IV. The scatter obviously arises from the fact that the mass of the planet(s) is not determined!…

Another problem remains: clearly it is out of the question to measure the intrinsic brightness of the planet. Still, its mass, as given in Table IV, remains quite high: it is of the order of ten times that of Jupiter. Under these conditions, is it really a planet?

It is after all known that many observed binary systems form highly disparate couples, involving a 'normal' star and a white dwarf. The white dwarfs have a special property: while in normal stars the central density depends on the temperature, in white dwarfs (as no doubt in planets) the central density depends on the pressure, because the matter is in a degenerate state: this has been known for 30 years, thanks to the work of Russell and Kothari.

If we attempt to go continuously from a stone to a white dwarf, the density is constant at first, and the radius increases as the cube root of the mass. At the scale of planets there are changes in the structure and the crystalline state of the matter due to pressure effects. For masses of the order of magnitude of Jupiter, an increase in the mass will result in a decrease in the radius, because of major changes in the physical properties. Will the mass-radius curve (Figure 39) join onto the white dwarf curve? Many theoreticians believe so, and the calculations of De Marcus for pure hydrogen masses, shown on Figure 39, suggest it.

Consequently, a determination of the mass is essential if we are to establish a distinction between planet and star. Now the intrinsic luminosity of a star is a function of the mass according to a fairly general relation:

$$\log L = 3.5 \log M. \tag{9}$$

Can such a relation be applied to planets? For an object of mass 0.05 $M_\odot$ we should have:
$$L = 2.5 \times 10^{-5} L_\odot, \tag{10}$$

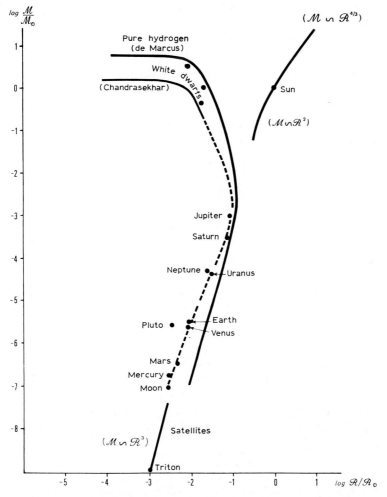

Fig. 39.   Relation between planetary masses and radii. Solid lines: predictions of the theory of stellar and planetary structure; dashed line: observations of planets and satellites.

and therefore:

$$\Delta m = -2.5 \log(2.5 \times 10^{-5}) = +11.5, \tag{11}$$

which would correspond to an intrinsic brightness definitely higher than the one we have calculated for the planet supposing it to be a perfect diffusor. In order to obtain an intrinsic brightness corresponding to a difference in brightness of at least 23 magnitudes (see above, p. 85), we would have to go down to masses lower than 0.002 $M_\odot$. This is not very far from Jupiter! The discussion is difficult: in fact it is doubtful that the mass-luminosity relation is valid; the luminosity should be lower than that predicted by this relation: moreover this is the case for the white dwarfs. It is in fact doubtful that the central temperature of a body of such low mass can be high enough for energy-producing nuclear reactions to occur. Thus the objects detected by Van de Kamp are most probably planets; the calculation carried out above is most probably inapplicable.

Another argument is often given in favor of empirical evidence of planetary systems. We know that the system of the star plus the planets possesses a certain amount of angular momentum. In the solar case, since the planets are distant from the sun and its speed of rotation is slow, practically all (98%) of the angular momentum is concentrated in the planets. On the contrary hot stars (hotter than those of spectral type F5) are observed to be rotating rapidly. The speed of rotation drops suddenly as one goes from hot stars to cooler stars; it is practically zero for stars cooler than those of type F5, and in particular for the sun. Now it is accepted that the angular momentum of the star-plus-planets system stays constant in the course of evolution. This leads us to think that if planets are formed, they take up the angular momentum of the system and have a braking effect on the rotation of the star. In truth this argument is quite indirect[*]; there are other theories of the rotation, which in particular consider coupling between rotation, meridional circulation, and convection in the star – all phenomena which strangely enough also appear for stars cooler than F5 but not for hotter ones; in this view rather than being absorbed by the planets, the angular momentum would be expelled in the ejection of matter accompanying strong activity analogous to solar activity.

In any case we shall retain Van de Kamp's determinations as an extremely strong argument in favor of the existence of planets revolving around stars of our Galaxy.

If this is indeed the case, we cannot help being struck by one fact: in our Galaxy, one star out of three is double, and 1 out of 30 is a white dwarf. Out of 39 nearby stars, 15 are binary systems, and at least 4 and no doubt more have planets. In addition, such an abundance of planets is to be expected: present cosmogenic theories fix their origin in the mass of dust surrounding stars in formation; this is certainly a very general phenomenon…

And so our solar system is not unique, but rather exists in millions of copies

---

[*] Moreover, if Dicke's assertion that the interior regions of the sun remain in rapid rotation is correct, the argument would become a bit more complicated, since those regions would then have conserved significant angular momentum.

throughout the universe. Conditions necessary for life (two planets, the earth and Mars, out of the 9 major planets of our solar system, certainly possess these) must no doubt exist on millions of earths – at least! And consequently it is not madness to consider the plurality of inhabited worlds as an extremely likely thing... And so we agree with Fontenelle!

The existence of planetary systems is thus extremely probable. However, on these planets, is *life* 'probable'? What about intelligent, highly evolved life, as on earth? In such a question (I shall come back to it later), it is extremely difficult to define a probability. In fact the length of time that life has existed on earth, as compared to the age of the earth, can hardly be considered a measure of the probability of the existence of life on a planet. To begin with, for how long will the phenomenon of life last on the earth? Who knows!... Moreover, even if we knew how to evaluate this lifetime, we would still not know the governing factors, so that we would hardly know how to generalize it to other worlds!

Still the question is worth looking into. The astronomer Su-Shu Huang believes that the important parameter governing the appearance and the maintenance of life is the quantity of energy E received (per unit surface and time) by a planet from its star.

Taking as our unit the quantity of energy received by the earth, we find the following values for each of the planets of the solar system:

Mercury	6.68	Mars	0.43
Venus	1.91	Jupiter	0.04
Earth	1.00	Saturn	0.01

We can conclude that life can appear and subsist only if E is greater than 0.1 and smaller than 5. We can consider in what region about a star the energy received lies within these limits, and how large this region is. Obviously this zone will be closer for fainter stars. It is easy to fix its extent in relative values. If D_1 and D_2 are the interior and exterior diameters of this region:

$$\frac{D_1}{D_2} = \sqrt{\frac{E_2}{E_1}} = \sqrt{\frac{0.1}{5}} \sim 0.14. \tag{12}$$

Clearly the geometrical extent of this 'zone of life' is very small for cold stars, so that the probability for these to have planets in which life may develop is no doubt rather low.

However, there is another point to be considered: the *lifetime of the star*. The evolution of life from its origin to its present state here on earth has required many centuries, millenia, indeed far longer. On the other hand, the evolution of some stars would not leave enough time for life to develop to a highly evolved state. The life history of a star can be summarized as follows: in the initial phase, a cold cloud of dust and gas condenses. Later, when thermonuclear reactions have exhausted the supply of hydrogen in the core of the star, it becomes a giant, and suddenly radiates far more: life and perhaps even the planets supporting it would probably disappear quite suddenly. The star continues to evolve, but its system can no longer support life. Now hot (or massive) stars go very quickly from the stage of condensation to the stage of thermonuclear conflagration; cold stars evolve much more slowly: one to ten million years for the most massive, up to one hundred billion (thousand million) for the coldest stars. The sun is about halfway. Statistics based on these considerations show that only 10% of the stars, the coolest, are favorable.

According to Su-Shu Huang, there is a third factor to be taken into consideration. This is the question whether the 'favorable' planets (i.e. those on which life might be possible) can have *stable orbits*, which will remain within the 'zone of the possibility of life'. It is believed that more than half (and perhaps much more than half) of all stars belong to binary or multiple systems. Among these stars, only 1–2% can have planets on which life is possible, for planetary orbits can easily be unstable in the case of multiple systems. Altogether, perhaps 'only' 3–5% of all stars may have planets which could be the abode of life, if we use the criteria of Huang... Still that leads to an enormous number, several billion, in our Galaxy alone.

Obviously, from the purely biological point of view, the evaluation of the probability that a planet – supposing the astronomical and astrophysical conditions to be favorable – will contain life, and even more so the evaluation of the probability that it is inhabited by evolved intelligent life, is an extremely risky affair!

Two schools of thought (and really it is most often intuition) affront one another on this point. One side believes the earth's situation to be exceptional. For them, if one could (but alas one cannot) obtain the statistics for all the planets in the Galaxy, from the biological point of view, it would turn out that our earth is on the 'tail' of the distribution, very far from the mean. The others believe that, on the contrary, the earth might be very much an 'average' planet in this imaginary survey. It is quite difficult to come to a conclusion! At least it can be said that on earth the appearance of life through physico-chemical processes could not have taken place very often: in fact life occurs in very well-defined forms, and the elementary amino acids which compose terrestrial living matter are very few in number: this argues in favor of the notion that life on earth is the result of an extremely rare and *a priori* highly improbable combination of circumstances. The opponents of this idea consider it exceedingly pessimistic, and cite the case of Mars and its vegetation (see above, pp. 79–83). However, this vegetation remains in controversy, since mineral substances can also have seasonal variations in appearance, if their crystallization water content varies...

Under these conditions, what is to be done, and what is to be believed? As we have seen, observation and theory are in agreement on the existence of many planets. Astronomy shows that a significant proportion of these *could* be inhabited (in the sense that they are 'habitable' – but without implying that it is very likely that they are inhabited). However, neither biological nor astronomical or physical theory gives us any information on the probability that life is indeed present. Given the difficulties we encounter now in observing such life on Mars and Venus, there is no chance of observing non-intelligent life on any other planet. Only intelligent life could find a way to survive some extremely long interstellar voyage or in some other way reveal its presence.

It is therefore understandable that the systematic search for such evidence has been proposed, and that attempts have been made to explain numerous otherwise inexplicable phenomena in terms of the existence of such intelligent life.

5. Project Ozma

The aim of Project Ozma (whose name was apparently chosen in honor of the wizard of Oz, the famous precursor of interstellar flight, by balloon, in the story by L. Frank Baum, well-known to younger generations for over 50 years) was indeed to look for possible signs of intelligence in the universe. It came out of the meeting of a theoretical idea due to Cocconi and Morrison and the techniques of radioastronomy in the hands of Frank Drake, applied to the planetary systems discovered by astronomers, which appear in Table IV (p. 86).

In fact the idea is quite simple. If there is life on another planet, the probability that it should be at the same stage as on earth is very small. Now mechanical or technological progress which follows biological evolution is rapid. Therefore it is likely (if life exists on many planets) that in many places in the universe life should be far more highly evolved, both biologically and technologically, than on earth. Consequently, these 'extraterrestrials' should be capable of communicating over very long distances, they might be aware of our existence, and could transmit powerful radio signals, invent universal languages (and ways to use them) etc....

Since they are intelligent, they will have chosen the most economical mode of communication: radio-wave transmission. They will have chosen a wavelength that it is sufficiently short not to be drowned out by galactic noise, but long enough not to be absorbed in the terrestrial atmosphere (Figure 40). Since they know physics

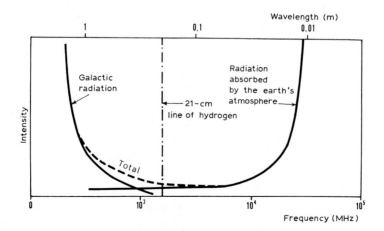

Fig. 40.   Choice of the wavelength for (hypothetical) interstellar radio communications.

and astrophysics, they know that the most interesting radiation for radioastronomy comes in the 21-cm line of neutral hydrogen, and they must suppose that we have radiotelescopes tuned in on this wavelength: they may indeed know that we are intelligent, or if they do not, they may in any case consciously avoid communicating with any beings so primitive as to be ignorant of radioastronomy. Therefore, if these intelligent beings exist, and if they want to converse with us, they will send us messages on the wavelength of 21 cm.

Now a distinct signal emitted by a planet at an intensity comparable to what we could emit from earth would be detectable up to a distance of 9 light-years by a radiotelescope of only 30 m in diameter: there are many such, and even some much larger! Now the systems of τ Cet and ε Eri (as well as those of α Cen and 61 Cyg, but their radiation would be hidden by the nearby radiation of the Milky Way) are closer to us than this distance. This is why for several years a radiotelescope at the National Radio Astronomical Observatory at Green Bank (West Virginia, U.S.A.) was used to observe these two stars over several years on the wavelength of 21 cm. After more than a half-century, an old dream of Charles Cros, who wanted to communicate with the Martians by focussing the sun's rays on their deserts, became reality: we attempted to communicate....

What were we looking for at Green Bank?

Obviously extraterrestrials know neither Morse code nor Esperanto... On the other hand, the laws of numbers and the equations of mathematics can be considered to be universal. Television techniques can also be used to transmit images... And so Drake's radiotelescope was looking for a type of 'purposive' signal, with periodic rhythms, over a period of months, before the attempt was abandoned!

After all the attempt had to be abandoned... For some this negative experiment was predictable, and they confirmed their faith in the solitude of the earth.

On the contrary, others thought that the extra-terrestrials must have given up such attempts centuries ago (Voltaire thought of it, but he had no radio telescope!) or better still prefer to use other means of communication, such as space probes visiting the solar system from time to time; the search for such extra-terrestrials on the earth itself has been proposed. Also one can envisage a search for infrared radiation in nearby planetary systems (the same as those accessible to Project Ozma), for such an analysis could lead to the detection of living matter, as it has suggested in the case of Mars. It should be underlined that this would not necessarily be intelligent life.

We should also add that while it is reasonable to think that intelligent beings at a level of evolution close to that of man might consider communicating with us, it is unlikely that civilizations at a much higher level would start a conversation. After all, we make very few attempts ourselves to 'communicate' with the bees, or with oysters.

Therefore the probability that these advanced civilizations will be interested in conversing with terrestrials is perhaps very low, and this may reduce still much further the probability that we will soon have clear proof of the existence of extraterrestrial, intelligent life.

6. The Improbable, The Probable, The Possible and Nonsense...

It could be expected that the sort of research we have been discussing, and which, as can be seen from the papers by Huang, Drake, McGowan and quite a few others, is quite serious, was certain to spark some much less serious conjectures...

There is the idea (due to the eminent astrophysicist Shklovsky) that Phobos, the satellite of Mars, is a hollow (and therefore artificial?) body... There is the explosion of June 30, 1908 in Siberia, said to be too powerful to be due to a meteorite fall (and thus due to the explosion of an extra-terrestrial device?). There is the radiosource CTA 102, which emits a variable and perhaps periodic signal (therefore produced by an extra-terrestrial intelligence?). Even the spectra of supernovae, civilizations which have disappeared, the 'evident' proof that the Incas possessed nuclear energy..., and who knows what else! are brought in. The list of extra-terrestrial interventions knows no end...

I would not think of making fun of this. Certainly we have to juggle hypotheses. Aren't they the necessary condition for any creative research? Still, a certain scientific accuracy, a certain amount of logical rigor are needed, the moreso when dealing with such a delicate field. Hypotheses provoke counter-hypotheses; without decisive experiments and in the absence of clear proof, the probabilities must be evaluated.

Phobos? Which one of the journalists who have become the ardent propagandists of an artificial Phobos has really attempted the discussion of the orbit of Phobos and the perturbations affecting it, from a celestial mechanics point of view? Siberia? The collision of a comet with the earth is an extremely more probable and more likely event than the tragic end of a visit by extra-terrestrials. CTA 102? All sorts of periodic oscillations in brightness exist in nature (e.g., the pulsating stars of the 'cepheid' type) and their explanation is purely physical: indeed the very existence of a period would rather tend to exclude an intelligent intervention! But we should add that even the existence of these fluctuations of CTA 102 is denied by the majority of radioastronomers. Why bother with the rest!

Still we must speak about it, for it is absolutely essential to condemn the exploitation of the public's curiosity for purely commercial ends. This exploitation is dangerous. It introduces a pretentious tendency to confuse what is possible, what is probable, what is improbable, what is false and what is true. Anything is 'possible' and so anything can exist! As Poincaré put it, more or less: "I am playing cards with a man who comes up with the king twelve times in a row. What is the probability that he is a professional cheat?" Looking at the so-called observations of flying saucers, the probability of psychopathological reactions or that of individual or collective hysteria (it won't have been the first time!) seems to me to be far higher than that of an extra-terrestrial visit; moreover it is known that those who are mentally ill are the hardest to convince of their illness. Furthermore it is not just a question of illness but also of hypersensitivity, affecting the interpretation of the retinal image, etc. Also, may I add that the existence of conscious fabricators, saucer swindlers, is a fact whose probability is far from being negligible. They are of the same variety as the cancer swindlers, who are sent to jail.

Faced with a difficult problem (which some like to call 'mystery'), such as that of supernovae, we must first look hard at physics. If we do not find a solution, we can begin to look elsewhere. Still the fact that the spectra of supernovae remain unexplained does not imply that they are of intelligent origin. Logic must be applied everywhere, and the 'hypothesis' of intelligence is here a clearcut swindle. Would it not be better to satisfy the curiosity of the public by giving examples of good scientific method (and project Ozma is not a bad example), rather than by ignoring logic and giving the most sensational solution as the truest, rather than defining (as was once done by a famous saucer advocate) truth as a property of what is pleasing, rather than placing fantasy on the level of science and science on the level of magic?

I hope I will be pardoned my vindictive pursuit of the moneychangers who clutter up our temple! Now let me go through a rapid summary:

Planets? Billions of them in our Galaxy alone, certainly.

Life on these planets? It may be possible on a hundredth of them; that does not mean it exists there. But it would be foolish to say that it doesn't.

Intelligent life on these planets? It has never been observed; it is not impossible; however, it is very unlikely that the earth appears to merit voyages of at least several years for highly evolved extra-terrestrials (unless it is an exceptional planet, but this is just the postulate we have rejected) – also it is very improbable that we shall succeed in communicating with them. Here again it would be foolish to say it is impossible. But after all, a very low probability is equivalent to an impossibility. If the monkeys in a zoo were given a set of typewriters (the example is due to Emile Borel) and suddenly became possessed of an urge to type without stopping, would they type exact copies of the books to be found in the Bibliothèque Nationale? This is not impossible. Still the probability is so low that no one would think for a moment that this possibility might some day become reality. We can say the same of many of the 'and even so, if?' that our technical progress leads us to ask about extra-terrestrials.

AFTER TEN YEARS OF SPACE RESEARCH

In 1968, more than 10 years after the launching of the first artificial satellite, 20 years after the first utilization of rockets as space observatories, it is tempting to ask to what extent space research will render obsolete the gigantic installations of the great observatories or radio-observatories established on the earth.

Already it is beyond doubt that the existence of space research (and especially its political, strategic, and even perhaps economic ramifications) has led to a striking evolution in the position of astronomy in the world. This science, archetype of disinterested research, at one time the refuge of solitary scientific workers (if not of maniacs and lunatics), gently humanized over the past decades by contact with physicists and their habits (teamwork, use of modern techniques), has now been contaminated by the contact with space research: secrecy – because of military impli-cations; gigantism – a necessary condition for space ventures; nationalism, even chauvinism, despite their being in principle so foreign to scientific work; the ever-present concern with efficiency and speed... (all of which are tendencies which I must say worry me more often than they seduce me), all these are growing significantly.

These tendencies take hold in the international organizations. All too often those working in space research, in this phase of rapid development, come to look upon the practitioners of traditional astronomy with some contempt or at least conde-scension, these low-priced telescopic observers, these unambitious theoreticians of stellar physics.

But isn't this true of all young techniques? The spectroscopists of the last century, the radioastronomers of 20 years ago, all depended on recent technical developments much more than the representatives of the traditional disciplines, and were oriented toward the methods more than toward the objects to be studied. They tended to be ignorant in astronomy, and to forget about the existence of their colleagues. Now that they have solved the initial technical problems, they have had the time to become interested in the true subject of their research: the study of the physical universe. The techniques they have mastered have become their slaves rather than their *raison d'être*. And often the encounter of old disciplines and new techniques has led to fruitful discoveries. Stellar evolution by way of the Hertzsprung-Russell diagram, many years ago, and more recently the discovery of quasars, are striking examples of this attitude of cooperation.

I am convinced that space astronomy will evolve in the same way, by progressive integration. Using space techniques, astronomers worthy of the name will obtain (indeed, this is already often the case!) information which will only have its full value

when combined with data from ground-based astronomy. On the other side, governments, either as rivals or as partners, will engage in a race in astronautical technology
which will involve astronomers very little – and this will not hurt them in any
way either.

The true problem which will then face the astronomical community will be a problem
of choices. In about a century from now, it is clear that an observatory on the moon
(or in orbit, or elsewhere) will be feasible, and will be more powerful and better – from
all points of view – than terrestrial observatories. However, it would I believe be folly
to consider that this extra-terrestrial observatory*, for all its merits, could completely
replace ground-based observatories. After all, life there will at the very least be very
peculiar if not difficult. Scientists will be able to go there for only very limited periods
of time, if only because of a psychological need for more natural surroundings (need
which we feel so much already in our automated metropolises, and which will be much
stronger there). The trips necessary back and forth, the energy needed for running
and maintaining the installation will make these operations very expensive. Of course,
highly elaborate automation will allow remote programming of the observations and
remote analysis of the observed data. There again the high price will make some
observations and measurements rather uneconomical.

This is no doubt the major reason which will force us to keep up our traditional
observatories and to use them; in fact it operates right now: in France we continue
to use Meudon or Nice, in spite of the more favorable altitude of the Pic-du-Midi,
even though it has been made comfortable and possesses high quality instruments…
and this would be true even if the instruments in question were still better!

There is another reason for keeping traditional instruments in service. They constitute in fact an excellent testing ground for new techniques; they are also an excellent
tool for teaching young astronomers. It seems probable to me that their marginal
utilization will continue to be necessary.

Faced with a clear research program on a well-defined subject, what will the
astronomer in 2068 do? Like the one of 1968, and the one of 1868, he will try to
'optimize' the problem. He will do what he can cheaply; he will accept to pay (by an
expedition to the mountains… or by using an automatic circumlunar station) only
if it is shown to be necessary. And that case may be fairly rare! Already today
astronomers sometimes look at the financial pit of space technology with some reti-
cence. A small part of the space funds would easily yield considerable progress on
the ground. While the astronomer must be present in the front-line groups, he needs
just as much to have solid research groups to work out astronomical problems on
the ground. A healthy balance must be maintained, and the first place among astronomical research techniques will not necessarily go to space research: once again let
us repeat that the present phase of discovery and exploration of a technical means
will be followed by the recognition of the unity of the science and by a concern with
optimizing the economics of research. There will be a place for the million astrono-

* See also the work *Les observatoires spatiaux* by the same author.

mers which can be predicted by a (rather risky) extrapolation of the demography of the profession between 1920 and 1966 to the year 2068... That will only be one astronomer per 1000 men, one hundred times more than in France today!

In any event, it is beyond doubt that the new space techniques have pushed far back certain recent frontiers of astronomical research.

Since this opening up of the field will develop further, what are the major ideas which astronomy will have produced in a century from now?

The history of astronomy is ancient. It began with a first phase from Antiquity to the middle of the 19th century: the phase of exploration, of *inventory* – what I will call the 'entomological'* phase. Moreover, the inventory is not closed, and this description of the immediate universe improves every day.

The second phase, set in motion perhaps by the discovery of solar activity and that of supernovae (and so, schematically, since Tycho and Galileo), is that in which we became aware of the *evolution* of stars, the evolution of the universe. Moreover, the problems of 'astrometaphysics', which are connected with the very foundation of all physics, arise at the same time. This phase of research is analogous to the study of the evolution of species by biologists; it has its specialized fields, its embryology, teratology, physiology... and it is also in full development.

The dominant aspect of research to come – the third phase – will be the arrival on the scene of biology as *directly* associated with astronomy: 'bioastronomy' is no doubt the new field which is destined to be developed in the most fundamental way – with as a corollary a detailed study of the planets, their exploration and indeed their utilization, for ends which will have nothing more to do so with astronomy.

However, the increasing power of rockets, the possibility of sending into orbit or to other planets heavier and heavier objects, increasingly automatic and accurate devices, will also carry the exploration of the universe in space and time very very far, Thus lowering the minimum observable brightness corresponds to a significant increase in the distance to which we can see: the most distant galaxies should help us to solve the problem of the expansion of the universe; measurements of the enormous energy output of the quasars and gamma-astronomy in general will lead us to an understanding of the most important evolutionary phenomena; closer to home, the study of the sun at short wavelengths (X-rays) will perhaps lead us to solve the mystery – excuse me, the problem – of its activity, and this will perhaps again lead us back to bio-astronomy and the problem of the origin of life. Of course such progress implies that along with the increasing power of space devices there is a healthy balance producing a parallel development of terrestrial techniques and especially teams of theoreticians and calculators.

The complete opening of the spectrum (especially toward short wavelengths), direct access to the moon and the planets (and therefore an effort towards the understanding of the bio-physical evolution of the universe), these constitute undeniable progress of obvious importance. We have one more technique, after astrometry, after spectro-

---

* Or taxonomic.

scopy, after radioastronomy,... at our service. May I express the hope that this new slave of man's ambitions should not become a rebel, or even enemy slave, and that the irresistible movement to the sky be organized in the respect of human solidarity on earth. May I then conclude this chapter and this book with the warning given by André Danjon, in the supplement to *Astronomie Populaire* devoted to extra-terrestrial devices:

We have just used the term *conquest*; it contains the best, it can hide the worst. Man will not carry with him his thirst for discovery and learning alone; he will not be able to liberate himself entirely of individual or group egotisms. For the honor and for the safety of the human species, this new step toward power must not become a cause for fear for those who aspire only to live in peace.

Following the enthusiasm provoked by the first successes, the echo of interested sentiments begins already to be heard. Thus, more and more in scientific congresses dealing with space, organizational questions must be dealt with in order to avoid the establishment of privileges. It is no longer a joke when one speaks of space law...

... Let us express the wish that scientific profit come before interested exploitation in the march towards space, and that we shall not talk too much of conquest; some of these are disastrous, and that of nuclear energy has not had only beneficial effects. This time all lovers of the beauties of the heavens are more or less involved in the event; may they bring with them a bit of their precious tradition which is to look upwards.

APPENDIX

TABLE V

Universal physical constants

Speed of light	$c = 2.99791 \times 10^{10}$ cm sec^{-1}
Gravitational constant	$G = 6.668 \times 10^{-8}$ dyne cm^2 g^{-1}
Planck's constant	$2\pi h = h = 6.6237 \times 10^{-27}$ erg sec
Charge of the electron	$e = 4.80217 \times 10^{-10}$ e.s.u.
Mass of the electron	$m_e = 9.1071 \times 10^{-28}$ g
Boltzmann constant	$k = 1.38024 \times 10^{-16}$ erg/deg
Avogadro's number (physical)	$\mathcal{N} = 6.0238 \times 10^{23}$
Gas constant	$\mathcal{R} = 8.3143 \times 10^7$ erg/deg/mole
Mass of the hydrogen ^{1}H atom	$M_1 = 1.6734 \times 10^{-24}$ g
Stefan's constant	$\sigma = \dfrac{8\pi^5 k^4}{15c^3 h^3}\dfrac{c}{4} = 5.6698 \times 10^{-5}$ erg cm^{-2} deg^{-4} sec^{-1}

Electron volt (1 eV):

associated wavelength	12396.3×10^{-8} cm
associated wave number	8067.1 cm^{-1}
associated frequency	2.41838×10^{14} sec^{-1}
associated energy	1.60184×10^{-12} erg
associated temperature	11605.9 K

Parsec	3.0857×10^{18} cm $= 206265$ AU $= 3.262$ light-years.
Astronomical unit (mean earth–sun distance)	AU $= 1.496 \times 10^{13}$ cm $= 499.01$ light-seconds
Light-year	l.y. $= 9.4605 \times 10^{17}$ cm $= 6.324 \times 10^4$ AU.

TABLE VI

The sun

Diameter: $2R = 1.3920 \times 10^{11}$ cm
Mass: $M = 1.989 \times 10^{33}$ g

Axial rotation (sidereal):

$\lambda = 0°$: $14°.5$ per day (equator)
$\lambda = 45°$: $13°.2$
$\lambda = 90°$: $11°.8$ (pole)

TABLE VII

The solar system

Planet	Satellites (non-exhaustive list)	Diameter (D) (km)	Mass (M) (in grams)	Semi-major axis of the orbit: a (in 10⁶ km)	Period of revolution (P) (in years or in days)	Period of axial rotation: D (days, hours)	Diverse remarks
☿ Mercury	(none)	4840	3.333×10^{26}	57.9	0.240yr	59d	No atmosphere
♀ Venus	(none)	12228	4.870×10^{27}	108.2	0.615yr	242.9d	Thick cloudy atmosphere
⊕ Earth	– / Moon	12742 / 3476	5.976×10^{27} / 7.35×10^{25}	149.6 / 0.3844	1.000yr / 27.321661d	23h 56m 4.09s / 27.3d	Atmosphere / No atmosphere
♂ Mars	– / Deimos / Phobos	6770 / (15) / (10)	6.443×10^{26} / ? / ?	227.9 / 0.02352 / 0.00937	1.881yr / 1.262d / 0.3189d	24h 37m 22.668s / – / –	Thin atmosphere (abundant CO_2)
♃ Jupiter	12 satellites / e.g.: Io / Europa / Ganymede / Callisto	140720 / 3550 / 3100 / 5600 / 5050	1.8993×10^{30} / 7.2×10^{25} / 4.7×10^{25} / 15.5×10^{25} / 9.7×10^{25}	778 / 0.422 / 0.671 / 1.070 / 1.880	11.862yr / 1.769d / 3.551d / 7.155d / 16.689d	9h 50m – 9h 56m / – / – / – / –	Thick atmosphere, polyatomic molecules: NH_3.
♄ Saturn	10 satellites / e.g.: Tethys / Dione / Titan	116820 / 1000 / – / 4950	5.684×10^{29} / 6.5×10^{23} / 1.0×10^{24} / 1.4×10^{26}	1427 / 0.295 / 0.377 / 1.222	29.456yr / 1.888d / 2.737d / 15.95d	10h 14m – 10h 40m / – / – / –	Thick atmosphere; Saturn has a very flat system of rings
♅ Uranus	5 satellites / e.g.: Ariel / Umbriel	47100 / – / –	8.676×10^{28} / – / –	2870 / 0.192 / 0.267	84.015yr / 2.520d / 4.144d	10h 45m / – / –	Atmosphere (?)
♆ Neptune	(2 satellites)	44600	1.029×10^{29}	4496	164.788yr	15.8h	
♇ Pluto	(none)	6000	5.53×10^{27}	(5900)	247.7yr	6.39d	Probably specularly reflecting surface.

Around the sun for the planets, around the planet for the satellites.

BIBLIOGRAPHIC ORIENTATION

There are few textbooks which deal with space research alone. On the other hand, a great many symposiums have dealt with the subject. We give a non-exhaustive list below.

Discussion on Observations of the Russian Artificial Earth-Satellites and their Analysis. Symposium of the Royal Society, London, 29 November 1957, Proc. Roy. Soc., ser. A, **248**, no. 1252, 1958.
Dynamique des satellites, IUTAM Symposium, Paris, 28–30 May 1962 (ed. by M. Roy), Springer, Berlin-Göttingen-Heidelberg, 1963.
Use of Artificial Satellites for Geodesy. 1st International Symposium, Washington, 26 April 1962 (ed. by G. Veis), North-Holland Publ. Co., Amsterdam, 1963.
Astronomical Observations from Space Vehicles. IAU Sympos. no. 23, Liège, August 1964 (ed. by J.-L. Steinberg), Publ. du C.N.R.S., Paris, 1965; *Ann. Astrophys.* **27** & **28** (1964–65).
Trajectories of Artificial Celestial Bodies as Determined from Observations. COSPAR-IAU-IUTAM Symposium, Paris, 20–23 April 1965 (ed. by J. Kovalevsky), Springer, Berlin-Heidelberg-New York, 1966.
International Space Science Symposia, organized by COSPAR and published under the title *Space Research*, by North-Holland Publ. Co., Amsterdam.
    I. *Nice, 1960* (ed. by H. Kallman), 1960.
    II. *Florence, 1961* (ed. by H. C. Van de Hulst, C. de Jager, and A. F. Moore), 1961.
    III. *Washington, 1962* (ed. by W. Priester), 1963.
    IV. *Warsaw, 1963* (ed. by P. Muller), 1964.
    V. *Florence, 1964* (ed. by D. G. King-Hele, P. Muller, and G. Righini), 1965.
    VI. *Mar del Plata, 1965* (ed. by R. L. Smith-Rose), 1966.
    VII. *Vienna, 1966* (ed. by R. L. Smith-Rose, with the collaboration of S. A. Bowhill and J. W. King), 2 vols, 1967.
    VIII. *London, 1967* (ed. by A. P. Mitra, L. G. Jacchia, and W. S. Newman), 1968.

Among *introductory* books on space research are the following:

R. E. Jastrow, *The Exploration of Space*, Macmillan, New York, 1960.
A. Danjon and P. Muller, 'Satellites artificiels et engins extra-terrestres', in *L'astronomie populaire*, Flammarion, Paris, 1960, chap. VIII.
A. I. Berman, *The Physical Principles of Astronautics*, Wiley, New York, 1961.
W. E. Liller (ed.), *Space Astrophysics*, McGraw-Hill, New York, 1961.
D. King-Hele, *Observing Earth Satellites*, Macmillan, London-Melbourne-Toronto, 1966.
Z. Kopal, *Telescopes in Space*, Faber & Faber, London, 1968.

The problems of *astronomy and astrophysics in general* are treated in a large number of books, at all levels (Translator's note: the French edition listed only books available in French. A list of books principally in English might include the following), going from the simplest to the most advanced:

J. C. Pecker, *The Sky*, The Orion Press, New York, 1960.
F. Hoyle, *Astronomy*, Doubleday, New York, 1962.
G. Abell, *The Exploration of the Universe*, 2nd edition, Holt, Rinehart & Winston, New York, 1969.
L. Motz and A. Duveen, *Essentials of Astronomy*, Wadsworth, Belmont, Calif., 1968.
A. Unsöld, *The New Cosmos*, Springer-Verlag, New York, 1968.
J. Kovalevsky, *Introduction to Celestial Mechanics*, Springer-Verlag, New York, and D. Reidel, Dordrecht, 1967.
J. Dufay, *Introduction to Astrophysics: the Stars*, Dover, New York, 1964.
T. L. Swihart, *Astrophysics and Stellar Astronomy*, Wiley, New York, 1968.

V. A. Ambartsumyan (ed.), *Theoretical Astrophysics*, Pergamon, London, 1958.
J. C. Pecker and E. Schatzman, *Astrophysique générale*, Masson, Paris, 1959.
A. Unsöld, *Physik der Sternatmosphären*, 2nd edition, Springer-Verlag, Berlin, 1955.

On the particular problem of the *planets*, the following works contain the essential bibliography:

Handbuch der Physik, vol. LII: *Astrophysik*; III: *The Solar System*, Springer, Berlin, 1959.
The Solar System. III: *Planets and Satellites* (ed. by G. P. Kuiper and B. M. Middlehurst), University of Chicago Press, 1961.
The Solar System. IV: *The Moon, Meteorites and Comets* (ed. by B. M. Middlehurst and G. P. Kuiper), University of Chicago Press, 1963.
J. Lequeux, *Planètes et satellites*, (Collection 'Que sais-je?', No. 383), P.U.F., Paris, 1964.
P. Guerin (ed.), *Planètes et satellites*, Larousse, Paris, 1967.

The possibility of extraterrestrial life has recently been treated in two very important and exhaustive books:

I. A. Shklovsky and C. Sagan, *Intelligent Life in the Universe,* Dell Publ. Co., New York, 1966.
E. U. Condon (ed.), *Scientific Study of Unidentified Flying Objects,* New York Times Book, Bantam Publ., New York, 1969.

Finally, there is an abundance of *journals* or *series* devoted to *space research*. The basic articles may be found in the following journals:

Advances in Astronautical Sciences, New York.
Annual Review of Astronomy and Astrophysics, Academic Press, New York-London.
Astronautica Acta, International Astronautical Federation, Springer, Vienna.
Planetary and Space Science, Pergamon Press, London-New York.
NASA Publications, Technical Notes, etc., Washington, D.C. (non-periodical).
Revue française d'astronautique, Paris.
La recherche spatiale, Centre National d'Etudes spatiales, Paris.

This bibliography may be completed by the following excellent sources of astronomical and astrophysical *data*:

Landolt-Börnstein, Gruppe VI: *Astronomy, Astrophysics and Space Research*, vol. I: *Astronomy and Astrophysics*, Springer, Berlin, 1965.
C. W. Allen, *Astrophysical Quantities*, 2nd edition, The Athlone Press, London, 1963.

Also by the regular *reports* of the following Commissions of the International Astronomical Union:

Commission 7 (Celestial Mechanics), Commission 16 (Physical Study of Planets and Satellites), Commission 17 (the Moon), Commission 44 (Astronomical Observations outside the Earth's Atmosphere).

INDEX OF NAMES

INDEX OF SUBJECTS

ASTROPHYSICS AND SPACE SCIENCE LIBRARY

Edited by

J. E. Blamont, R. L. F. Boyd, L. Goldberg, C. de Jager, Z. Kopal, G. H. Ludwig, R. Lüst,
B. M. McCormac, H. E. Newell, L. I. Sedov, Z. Švestka, and W. de Graaff

1. C. de Jager (ed.), *The Solar Spectrum. Proceedings of the Symposium held at the University of Utrecht, 26–31 August, 1963.* 1965, XIV + 417 pp.   Dfl. 50.—

2. J. Ortner and H. Maseland (eds.), *Introduction to Solar Terrestrial Relations. Proceedings of the Summer School in Space Physics held in Alpbach, Austria, July 15–August 10, 1963 and Organized by the European Preparatory Commission for Space Research.* 1965, IX + 506 pp.   Dfl. 65.—

3. C. C. Chang and S. S. Huang (eds.), *Proceedings of the Plasma Space Science Symposium, Held at the Catholic University of America, Washington, D.C., June 11–14, 1963.* 1965, IX + 377 pp.
                                                                                  Dfl. 68.—

4. Zdeněk Kopal, *An Introduction to the Study of the Moon.* 1966, XII + 464 pp.   Dfl. 72.—

5. Billy M. McCormac (ed.), *Radiation Trapped in the Earth's Magnetic Field. Proceedings of the Advanced Study Institute, Held at the Chr. Michelsen Institute, Bergen, Norway, August 16– September 3, 1965.* 1966, XII + 901 pp.   Dfl. 130.—

6. A. B. Underhill, *The Early Type Stars.* 1966, XIII + 282 pp.   Dfl. 56.—

7. Jean Kovalevsky, *Introduction to Celestial Mechanics.* 1967, VIII + 427 pp.   Dfl. 30.—

8. Zdeněk Kopal and Constantine L. Goudas (eds.), *Measure of the Moon. Proceedings of the Second International Conference on Selenodesy and Lunar Topography held in the University of Manchester, England, May 30–June 4, 1966.* 1967, XVIII + 479 pp.   Dfl. 90.—

9. J. G. Emming (ed.), *Electromagnetic Radiation in Space. Proceedings of the Third ESRO Summer School in Space Physics, held in Alpbach, Austria, from 19 July to 13 August, 1965.* 1968, VIII + 307 pp.   Dfl. 58.—

10. R. L. Carovillano, John F. McClay, and Henry R. Radoski (eds.), *Physics of the Magnetosphere. Based upon the Proceedings of the Conference held at Boston College, June 19–28, 1967.* 1968, X + 686 pp.   Dfl. 130.—

11. Syun-Ichi Akasofu, *Polar and Magnetospheric Substorms.* 1968, XVIII + 280 pp.   Dfl. 55.—

12. Peter M. Millman (ed.), *Meteorite Research. Proceedings of a Symposium on Meteorite Research held in Vienna, Austria, 7–13 August, 1968.* 1969, XV + 941 pp.   Dfl. 160.—

13. Margherita Hack (ed.), *Mass Loss from Stars. Proceedings of the Second Trieste Colloquium on Astrophysics, 12–17 September, 1968.* 1969, XII + 345 pp.   Dfl. 65.—

14. N. D'Angelo (ed.), *Low-Frequency Waves and Irregularities in the Ionosphere. Proceedings of the 2nd ESRIN-ESLAB Symposium, held in Frascati, Italy, 23–27 September, 1968.* 1969, VII + 218 pp.   Dfl. 43.—

15. G. A. Partel (ed.), *Space Engineering. Proceedings of the Second International Conference on Space Engineering, held at the Fondazione Giorgio Cini, Isola di San Giorgio, Venice, Italy, May 7–10, 1969.* 1970, XI + 728 pp.   Dfl. 140.—

p.t.o.

16. S. Fred Singer (ed.), *Manned Laboratories in Space. Second International Orbital Laboratory Symposium.* 1969, XIII + 133 pp.   Dfl. 30.—

17. B. M. McCormac (ed.), *Particles and Fields in the Magnetosphere. Symposium Organized by the Summer Advanced Study Institute, held at the University of California, Santa Barbara, Calif., August 4–15, 1969.* 1970, XI + 450 pp.   Dfl. 85.—

18. V. Manno and D. E. Page (eds.), *Intercorrelated Satellite Observations related to Solar Events. Proceedings of the Third ESLAB/ESRIN Symposium held in Noordwijk, The Netherlands, September 16–19, 1969.* 1970. XVI + 627 pp.   Dfl. 115.—

SOLE DISTRIBUTORS FOR U.S.A. AND CANADA:

Vols. 2–6, and 8: Gordon & Breach Inc., 150 Fifth Ave., New York, N.Y. 10011

Vols. 7 and 9ff.: Springer-Verlag New York, Inc., 175 Fifth Ave., New York, N.Y. 10011